This Book is the Property of:
Hichborn Consulting Group

Performance Criteria for Concrete Durability

OTHER BOOKS ON CONCRETE AND CONSTRUCTION
MATERIALS AVAILABLE FROM E & FN SPON

Application of Admixtures in Concrete
Edited by A.M. Paillere

Autoclaved Aerated Concrete: Properties, Testing and Design
RILEM Technical Committees 78-MCA and 51-ALC

Cement-based Composites: Materials, Mechanical Properties and Performance
A.M. Brandt

Concrete in the Marine Environment
P.K. Mehta

Concrete Mix Design, Quality Control and Specification
K.W. Day

Construction Materials: Their Nature and Behaviour
Edited by J.M. Illston

Creep and Shrinkage of Concrete
Edited by Z.P. Bažant and I. Carol

Durability of Concrete in Cold Climates
M. Pigeon and R. Pleau

Euro-Cements: Impact of ENV 197 on Concrete Construction
Edited by R.K. Dhir and M.R. Jones

Ferrocement
Edited by P. Nedwell and R.N. Swamy

High Performance Concrete: From Material to Structure
Edited by Y. Malier

Hydration and Setting of Cements
Edited by A. Nonat and J.C. Mutin

Manual of Ready-Mixed Concrete
J.D. Dewar and R. Anderon

Special Concretes: Workability and Mixing
Edited by P.J.M. Bartos

Thermal Cracking in Concrete at Early Ages
Edited by R. Springenschmid

For details of these and other books, contact The Promotions Department,
E & FN Spon, 2–6 Boundary Row, London SE1 8HN, UK. Tel: 0171–865
0066.

RILEM REPORT 12

Performance Criteria for Concrete Durability

State of the Art Report prepared by
RILEM Technical Committee TC 116–PCD,
Performance of Concrete as a Criterion
of its Durability

Edited by
J. Kropp

Labor für Baustofftechnologie, Hochschule Bremen, Germany

and

H.K. Hilsdorf

Institut für Massivbau und Baustofftechnologie Universität
Karlsruhe, Germany

E & FN SPON
An Imprint of Chapman & Hall

London · Glasgow · Weinheim · New York · Tokyo · Melbourne · Madras

**Published by E & FN Spon, an imprint of Chapman & Hall,
2–6 Boundary Row, London SE1 8HN**

Chapman & Hall, 2–6 Boundary Row, London SE1 8HN, UK

Blackie Academic & Professional, Wester Cleddens Road, Bishopbriggs,
Glasgow G64 2NZ, UK

Chapman & Hall GmbH, Pappelallee 3, 69469 Weinheim, Germany

Chapman & Hall USA, 115 5th Avenue, New York NY 10003, USA

Chapman & Hall Japan, ITP-Japan, Kyowa Building, 3F, 2-2-1
Hirakawacho, Chiyoda-ku, Tokyo 102, Japan

Chapman & Hall Australia, 102 Dodds Street, South Melbourne, Victoria
3205, Australia

Chapman & Hall India, R. Seshadri, 32 Second Main Road, CIT East
Madras 600 035, India

First edition 1995

© 1995 RILEM

Typeset in 10/12 pt Times by Florencetype Ltd, Stoodleigh, Tiverton,
Devon
Printed in Great Britain by T.J. Press (Padstow) Ltd, Padstow, Cornwall

ISBN 0 419 19880 6

A catalogue record for this book is available from the British Library

∞ Printed on acid-free text paper, manufactured in accordance with
ANSI/NISO Z39,48-1992 (Permanence of Paper).

Contents

RILEM technical committee 116-PCD and contributors

Professor Dr. -Ing. H.K. Hilsdorf *(Chairman 1989–1992)*
Institut für Massivbau und Baustofftechnologie, Universität Karlsruhe, Kaiserstr. 12, D-76131 Karlsruhe 1, Germany

Professor Dr. -Ing. J. Kropp *(Chairman 1992–1995)*
Labor für Baustofftechnologie, Hochschule Bremen Neustadtswall 30, D-28199 Bremen, Germany

J. Aldred
CEMENTAID (S.E.A.) Pte Ltd, 12 Neythal Road, RS Singapore 2262, Republic of Singapore

Professor em. Dr sc. techn. Altner
Universität Leipzig, Karl-Liebknecht-Str. 132, D-04275 Leipzig, Germany

Dr C. Andrade
Instituto Eduardo Torroja de la Construccion y del Cemento, Serrano Galvache s/n, Apartado 19002, E-28033 Madrid, Spain

Dr N.R. Buenfeld
Department of Civil Engineering, Imperial College, Imperial College Road, London SW7 2BU, United Kingdom

Dr N.J. Carino
National Institute of Standards and Technology, Gaithersburg, MD 20899, USA

Dipl. -Min. F. Ehrenberg
Institut für Baustoffe, Massivbau und Brandschutz, TU Braunschweig, Beethovenstr. 52, D-38106 Braunschweig, Germany

Dr E.J. Garboczi
Buiding Materials Division, National Institute of Standards and Technology, Gaithersburg, MD 20899, USA

Dr M. Geiker
COWIconsult, Consulting Engineers, Parallelvej 15, DK-2800 Lyngby, Denmark

Professor Dr O.E. Gjørv
University of Trondheim, Department of Civil Engineering, Division of Building Materials, N-7034 Trondheim NTH, Norway

Dr F. Goncalves
LNEC, Dep de Materiais de Construcao, Avenida do Brasil, 101, P-1799 Lisbao Codex, Portugal

Dr Ing. H. Grube
Forschungsinstitut der Zementindustrie, Tannenstr. 2, D-40476, Düsseldorf, Germany

Professor R.D. Hooton	University of Toronto, Department of Civil Engineering, Toronto, Ontario M5S 1A4, Canada
Dr M. Massat	INSA, UPS Genie Civil, 156 Avenue de Rangueil, F-31077 Toulouse Cedex, France
Dr -Ing. S. Modry	Technical University of Prague, Building Research Institute, Solinova 7, CS-16608 Praha 6, Czech Republic
Dr C. Molin	BARAB, Tullgardsvägen 12, Box 4909, S-11694 Stockholm, Sweden
Professor Dr L.O. Nilsson	Chalmers University of Technology, Division of Building Materials, S-412 96 Göteborg, Sweden
Professor Dr J.-P. Ollivier	INSA, UPS Genie Civil, 156 Avenue de Rangueil, F-31077 Toulouse Cedex, France
Professor C.L. Page	Department of Civil Engineering, Aston University, Aston, Birmingham B4 7ET, United Kingdom
Dr L.J. Parrott	British Cement Association, Century House, Telford Avenue, Crowthorne, Berkshire RG11 6YS, United Kingdom
Dipl. Ing. K. Paulmann	Institut für Baustoffe, Massivbau und Braunschweig, TU Braunschweig, Beethovenstr. 52, D-38106 Braunschweig, Germany
Dr P.E. Petersson	Swedish National Testing and Research Institute, Building Technology, PO Box 857, S-501 15 Boraas, Sweden
Dr L. Pozsgai	YBL Miklos Epitöipari, Mpszaki Föiskola, Postfach 117, H-1442 Budapest, Hungary
Mr F.R. Rodriguez	Lab Central de Estructuras y Materiales, CEDEX, c/Alfonso XII, No 3, E-28014 Madrid, Spain
Dr J.P. Skalny	11910 Thurloe Drive, Timonium, MD 21093, USA
Dr Tang Luping	Chalmers University of Technology, Division of Building Materials, S-412 96 Göteborg, Sweden
Dr R. Torrent	Holderbank, Management und Beratung AG, CH-5113 Holderbank, Switzerland
Dr D. Whiting	Construction Technology Laboratories, 5420 Old Orchard Road, Skokie, IL 60077, USA

Preface

During recent years the durability of concrete structures has attracted considerable interest in concrete practice, materials research, and national as well as international standardization. Owing to extensive efforts much progress has been made to understand better the corrosive mechanisms acting on concrete and reinforced concrete structures in service, and frequently, more restrictive regulations have been introduced to aim at an improved durability of structures. These rules are essentially based on long-term experiences rather than on a designed corrosion resistance. Although the basic mechanisms of the most important corrosive actions are well known, no generally accepted methods are available yet to predict the performance of a given concrete on the basis of a materials characteristic in relation to the severity of the corrosive attack that is expected for the prevailing exposure condition.

The permeability of concrete for various agents has been considered to represent such a materials parameter, which could serve as a performance criterion for concrete durability.

In 1989 the RILEM Technical Committee 116-PCD *Permeability of concrete as a criterion of its durability* was established upon the initiative of Professor Dr Hilsdorf, who also chaired the committee until 1992. The objectives of TC 116-PCD were to investigate the potentials of this approach, to monitor and evaluate international efforts on testing concrete for its potential durability and, if the concept appeared promising, provide recommendations on suitable test methods to determine the relevant concrete characteristics. The possible applications for corresponding test methods and the emerging results could range from routine quality control to sophisticated calculation models for corrosion rates and service life prediction for structures. Numerous technical committees and working groups within RILEM as well as other international organizations are engaged in related tasks.

Experts from many countries joined the RILEM Technical Committee 116-PCD as full members as well as corresponding members to discuss and contribute ideas or send in their written comments on a number of sub-tasks, which were established on the objectives of the committee. One basic task was to compile a state-of-the-art report on the importance of transport parameters, their measurement, and verified correlations with concrete durability. The literature survey should allow a critical evaluation of the approach, thereby also considering alternative concepts. *Performance Criteria for Concrete Durability* summarizes the reviewed literature and, in conclusion, supports the concept.

The outline of the major topics in this report emerged from the lively cooperation in the Technical Committee, and numerous committee members volunteered to prepare the individual chapters of the report. Their draft chapters have been intensively discussed in the committee meetings held at Karlsruhe, Düsseldorf, Birkerod, Toulouse, Madrid, Gotheborg and Lisbon, and valuable hints, criticisms and assistance in reviewing the literature were extended by all the committee members to each author. Nevertheless, the chapters as presented in this report represent the personal views of the respective authors. However, the conclusions drawn from each chapter had to find acceptance, and therefore they are supported by the whole committee.

The preparation of a state-of-the-art report involves an ongoing temptation to update to account for the latest developments and results obtained by the authors themselves or those published in the literature. However, this in turn will prevent finalizing of the report, especially if various aspects of the subject are so widely dealt with as given here. The authors as well as the committee have been aware of this problem! In this respect it is important to recall that the report serves as a basis for the committee's further work, which, in the second step, concentrates on the development of test methods; only then will guidelines on test procedures be drafted. Aside from the further results that will evolve from this experimental work within the committee, of course ongoing research outside as well as more recent publications that were not considered for editorial reasons will then also be taken into account.

Until stepping down from the chairmanship, Professor Dr Hilsdorf guided and directed the committee work, and I would like to take this opportunity and thank him – and also on behalf of the whole committee – for sharing his knowledge and experience with us. I also wish to thank all the authors for their efforts to prepare the chapters, their cooperation and patience to support requests for editorial amendments, and equally, thanks to the full and corresponding committee members who promoted the work by their discussion contributions, criticisms and proposals.

And again it is the merit of Mania D Siggelkow that a large number of unique contributions could be shaped into a uniform report. Thank you Mania for your enjoyable cooperation.

Jörg Kropp
Chairman TC 116-PCD
Bremen

1 Introduction and problem statement

Hubert K. Hilsdorf

At present, the durability of concrete is controlled almost exclusively by specifying certain requirements for concrete composition, properties and composition of concrete constituents, casting and compaction procedures, curing and sometimes concrete compressive strength. This approach frequently yields unsatisfactory results, and it is a common objective of concrete researchers and engineers to develop performance criteria that would allow more reliable estimates of the potential durability of a given concrete mix and of the probable durability of a concrete structure [1.1].

In many codes and specifications the standard compressive strength of concrete is used as such a criterion. Though it may give some indication of the potential durability of concrete, such as its resistance to carbonation, it cannot be a generally valid criterion for a number of reasons, one of them being that the standard compressive strength of a concrete cube or cylinder constitutes the mean value of a property of an entire cross-section, whereas concrete durability is governed primarily by the properties of the concrete skin. Better correlations between durability and strength properties may be achieved if the strength of concrete at the end of curing is used as a reference since at this stage the strength of a concrete specimen may reflect more closely the properties of the concrete skin at a later age [1.2].

It is generally accepted that concrete durability is to a large extent governed by concrete's resistance to the penetration of aggressive media. Therefore, a criterion that is based upon such resistance should be the most reliable approach.

In the search for such a criterion it has to be kept in mind that aggressive media may be present in a liquid or gaseous state and that they may be transported by various mechanisms: in particular permeation, diffusion, absorption and capillary suction, and combinations thereof. It is, therefore, at least questionable whether a single property will be sufficient to describe the resistance of concrete to the penetration of various types of aggressive medium. In addition, little experience exists with test methods

Performance Criteria for Concrete Durability, edited by J. Kropp and H.K. Hilsdorf. Published in 1995 by E & FN Spon, London. ISBN 0 419 19880 6.

that may allow the measurement of such resistance. Nevertheless, some previous experience has shown that correlations exist between material parameters controlling different transport mechanisms, such as diffusion coefficients and permeability coefficients, and also between transport characteristics and concrete durability characteristics, such as rate of carbonation. Therefore, the first objective of the work of RILEM Technical Committee TC 116-PCD was to prepare a state-of-the-art report dealing with the following subjects:

- transport mechanisms and their analytical description;
- relations between different transport mechanisms;
- relations between transport parameters and concrete durability characteristics;
- influence of concrete parameters on transport characteristics;
- influence of environmental parameters and moisture on transport characteristics,
- test methods to determine transport characteristics.

This state-of-the-art report will serve as a basis for further activities of RILEM TC 116-PCD, in particular:

- round-robin tests on test methods to measure concrete permeability,
- development of a recommended test procedure to measure concrete permeability,
- provision of examples of permeability values for a practical range of concretes; and
- indication of typical associated durability data.

In this report particular emphasis is placed on concrete permeability. However, depending on the outcome of the initial studies, concrete permeability may be replaced by other transport characteristics or combinations thereof.

Unless stated otherwise in the individual chapters the considerations presented in this report focus on normal structural concretes with a dense structure built up by a continuous cement paste matrix, normal weight aggregates and a compressive strength of the concretes in an intermediate range. The results reported may not necessarily apply to concretes with a special composition or structure, e.g. lightweight concrete with a honeycombed structure, use of special aggregates or additions such as polymers, or having very high compressive strengths. For such materials the durability may be controlled by additional mechanisms. Furthermore, the test methods described in this report were originally designed for normal concretes. Their suitability for testing special concretes is uncertain.

The experimental results reported on concrete transport parameters may significantly depend on the test procedure and the test conditions chosen in the respective investigation. For further use of these transport coefficients, e.g. as input data in numerical calculations or prediction models, reference should be made to the original publication in order to check on the test conditions that lead to the reported values.

References

1.1.   Hilsdorf, H.K. (1989) Durability of concrete – a measurable quantity? in *Proceedings of IABSE Symposium*, Lisbon, Vol. 1, pp. 111–123.
1.2.   Parrott, L.J. (1990) Carbonation, corrosion and standardization, in *Protection of Concrete*, Proceedings of International Conference, Dundee, Scotland, September, Eds R.K. Dhir and M.R. Jones, E & FN Spon, London, pp. 1009–1023.

2 Transport mechanisms and definitions

Jörg Kropp, Hubert K. Hilsdorf, Horst Grube,
Carmen Andrade and Lars-Olof Nilsson

The ingress of gases, water or ions in aqueous solutions into concrete takes place through pore spaces in the cement paste matrix or micro-cracks. A variety of different physical and/or chemical mechanisms may govern the transport of these media into the concrete, depending on the substance flowing and its local concentration, the environmental conditions, the pore structure of concrete, the pore radius or width of micro-cracks, the degree of saturation of the pore system and the temperature. Considering the wide range of pore sizes and a varying moisture concentration in the concrete as a function of the climatic exposure conditions, the transport of media into concrete in most cases is not due to one single mechanism. but several mechanisms may act simultaneously.

In the experimental investigation of the transport characteristics of concrete an attempt is normally made to limit the flow of media to one single transport mechanism in order to derive transport coefficients according to the underlying theoretical model of the transport process. This procedure, however, is subject to limitations, because in most cases concrete as the penetrated material does not behave as an inert and stable solid with uniform pores. As a consequence the physical structure of concrete may alter, chemical absorption may occur, and a multiple transport mechanism may take place during a test. Therefore, simplifying assumptions must be taken into account, and standard test procedures are mandatory in order to attain a sufficient reproducibility of the results.

In the following sections those transport mechanisms are presented that are normally associated with the ingress of deleterious material into concrete. The transport mechanisms also serve as the theoretical models to evaluate the transport characteristics of concrete in different test methods.

Performance Criteria for Concrete Durability, edited by J. Kropp and H.K. Hilsdorf. Published in 1995 by E & FN Spon, London. ISBN 0 419 19880 6.

2.1 Diffusion

Transfer of mass by random motion of free molecules or ions in the pore solution resulting in a net flow from regions of higher concentration to regions of lower concentration of the diffusing substance.

The rate of transfer of mass through unit area of a section, F,

$$F = \frac{dm}{dt}\frac{1}{A}$$

(2.1)

where F = mass flux (g/m²s)
m = mass of substance flowing (g)
t = time (s)
A = area (m²)

is proportional to the concentration gradient dc/dx and the diffusion co-efficient D (m²/s).

This relation is expressed in **Fick's first law** of diffusion [2.1]:

$$F = -D\frac{dc}{dx}$$

(2.2)

where D = diffusion coefficient (m²/s)
c = concentration (g/m³)
x = distance (m)

2.1.1 DIFFUSION COEFFICIENT

For solids, the characteristic material property D describes the ability of transfer for a given substance.

In many cases, the transfer of mass is superimposed by mechanisms other than random motion of molecules such as surface flow of condensed gas along pore walls or saturated capillary flow due to surface forces. In porous solids, for example, moisture may flow as the diffusion of water vapour; at the same time non-saturated or even saturated capillary flow may occur in finer pores [2.2]. Although additional transport mechanisms are acting in these cases, Fick's law of diffusion may be applied to quantify the multiple transport phenomenon [2.3]. A concentration gradient is then considered as the common driving force. The diffusion coefficient D, however, may depend strongly on the local concentration c of free ions or molecules. In inhomogeneous solids D may depend on location x.

Additionally, for ageing materials, D depends on time t, and in all cases diffusion is influenced by temperature T [2.4]. Therefore, the diffusion coefficient is a function of different variables:

$$D = f(c, x, t, T)$$

For transient diffusion processes the balance equation (2.3) describes the change of concentration in a unit volume with time. This equation is referred to as **Fick's second law** of diffusion [2.5]:

$$\frac{\partial c}{\partial t} = \frac{\partial}{\partial x}\left(D\frac{\partial c}{\partial x}\right)$$

(2.3)

with D being constant or a function of different variables.

For simple cases with respect to geometry and boundary conditions as well as for the diffusion coefficient D being constant, solutions exist for eq. (2.3) [2.6].

If the boundary condition is specified as $c = c_0$ and the initial condition is specified as $c = 0$ for $x > 0$, $t = 0$, a solution of eq. (2.3) is given by

$$c = c_0\left[1 - erf\left(\frac{x}{2\sqrt{Dt}}\right)\right]$$

(2.4)

where erf is the error function. Values of erf $(x/2\sqrt{(Dt)})$ vs $x/2\sqrt{(Dt)})$ are available either in mathematical tables or may be calculated with the help of a computer [2.5]. If experimental data of c vs x at time t are known, the diffusion coefficient D can be determined from eq. (2.4) by successive approximation to give the best fit, e.g. according to the method of least squares.

The diffusing substance may be partially immobilized due to chemical interaction or physical adsorption due to mass forces. Then the balance equation must be extended by a sink s [2.5]:

$$\frac{\partial c}{\partial t} = \frac{\partial}{\partial x}\left(D\frac{\partial c}{\partial x}\right) - s$$

(2.5)

The binding capacity of the penetrated material may be a variable of different parameters. It may depend on the local concentration of diffusing substance, as is the case for chlorides dissolved in the pore water. Furthermore, temperature effects may occur.

Because the binding capacity is governed by the chemical composition and the pore structure of the penetrated solids, the binding capacity may alter if changes of the penetrated material occur, e.g. due to carbonation. Therefore, the sink s is a function of different variables.

Immobilization of diffusing material is of considerable importance for the experimental determination of diffusion coefficients. As long as the binding capacity of the test specimen is not yet exhausted, the net flow of diffusing material appears to be low. Then, the diffusion coefficient is underestimated. This will result in erroneous estimates when applying eq. (2.2) or (2.3). In cases where the sink s is not taken into account an 'apparent' diffusion coefficient may be deduced from experiments, which then depends on time t.

For reinforced concrete structures, ingress of chlorides is one of the most important transport problems associated with immobilization of the diffusing chloride ion. Therefore, this mechanism will be presented in more detail in Chapter 6.

2.2 Permeation

Flow of liquids or gases caused by a pressure head.

Depending on the pore structure of the solid and the viscosity of the flowing media, capillary flow may be **laminar** or **turbulent**. For a turbulent flow the volume transported is not proportional to the pressure head.

2.2.1 COEFFICIENT OF PERMEABILITY

The coefficient of permeability is a materials characteristic describing the permeation of gases or liquids through a porous material due to a pressure head. The coefficient of permeability is determined experimentally and usually a laminar flow in the pore system is assumed [2.7].

2.2.2 GAS PERMEABILITY

For gases, the compressibility as well as the viscosity of the gas must be considered in order to derive the coefficient of permeability (m²) [2.8]:

$$K_g = \eta \frac{Ql}{tA} \frac{2p}{(p_1 - p_2)(p_1 + p_2)}$$

(2.6)

where
K_g = coefficient of permeability (m²)
η = viscosity of the gas (Ns/m²)
Q = volume of gas flowing (m³)
l = thickness of penetrated section (m)
A = penetrated area (m²)
p = pressure at which volume Q is measured (N/m²)
p_1 = pressure at entry of gas (N/m²)
p_2 = pressure at exit of gas (N/m²)
t = time (s)

Equation (2.6) is strictly valid only for laminar flow. This condition does not necessarily prevail for the transport of gases through the pore system of the hydrated cement paste because of the wide range of pore diameters. Therefore, the coefficient of gas permeability has also been defined according to eq. (2.7). In this case, K_g is no longer a materials characteristic but depends on the transported medium [2.9]:

$$\overline{K}_g = \frac{Q}{t}\frac{l}{A}\frac{p}{p_1 - p_2}$$

(2.7)

where $\overline{K}_g$ = coefficient of permeability (m²/s)
    Nevertheless, in this state-of-the-art report a coefficient of gas permeability as defined by eq. (2.6) is used.

2.2.3 WATER PERMEABILITY

Among liquids penetrating through concrete, water represents the most important fluid. In contrast to gases, liquids may be considered as incompressible, and eq. (2.6) may then be rewritten. If the viscosity of the liquid is taken into consideration and a laminar flow is assumed, the **coefficient of water permeability** is given by

$$K_w = \frac{Q}{t}\frac{l}{A}\frac{\eta}{\Delta p}$$

(2.8)

where K_w = coefficient of permeability (m²)
    Q = volume of liquid flowing (m³)
    t = time (s)
    l = thickness of penetrated section (m)
    A = penetrated area (m²)
    η = viscosity (Ns/m²)
    p = pressure difference (N/m²)

The coefficient of permeability represents a materials characteristic and is independent of the properties of the liquid.
    Frequently, the flow of water is evaluated according to the empirical formula of **D'Arcy** [2.10]:

$$K_w^* = \frac{Q}{t}\frac{l}{A}\frac{1}{\Delta h}$$

(2.9)

where K_w^* = coefficient of water permeability (m/s)
    Δh = pressure head (= height of water column) (m)

Then, K_w^* is not a materials characteristic but only describes the flow of water.
    For conversion of the coefficient of water permeability according to D'Arcy into the coefficient of water permeability according to eq. (2.8)

the viscosity of water has to be introduced and the pressure head given in terms of height of water column must be expressed in N/m² [2.11]. This results in a conversion factor of

$$\frac{\eta}{\rho g}$$

where η = viscosity of water (Ns/m²)
ρ = density of water (kg/m³)
g = gravity (m/s²)

or, for $T = 20°C$:

$$K_w^* = 9.75 \times 10^6 K_w \qquad (2.10)$$

It must be considered, however, that the viscosity of water is affected by temperature.

2.3 Capillary suction

Transport of liquids in porous solids due to surface tension acting in capillaries [2.12].
   The transport of the liquid is influenced by the following characteristics of the liquid:

- viscosity   (Ns/m²)
- density   (kg/m³)
- surface tension   (N/m)

and by characteristics of the solid, such as pore structure (radius, tortuosity and continuity of capillaries), and surface energy.
   The flow F due to **steady-state capillary suction** of water in a pore system is given by D'Arcy's law modified for non-saturated flow of water:

$$F = -\frac{k_p}{\eta}\frac{dp_w}{dx} \qquad (kg/m^2s)$$

$$(2.11)$$

where dp_w/dx is the gradient of pore water pressure p_w (N/m²) and k_p is the coefficient of moisture permeability (kg/m).
   The pore water pressure p_w is given by the air pressure p_a, the radius of the water meniscus r_m and the surface tension according to the **Laplace equation** [2.13]:

$$p_a - p_w = \frac{2\sigma}{r_m}$$

$$(2.12)$$

When only small pores are filled with water, for instance because of continuous drying, the pressure is very low ($p_w \ll 0$). This causes capillary suction to those areas, provided there is a continuous liquid path.

The pore water pressure p_w is also given by the water vapour pressure at saturation, p_s, and the relative humidity φ in the pore:

$$p_w = p_s + \frac{RT\rho}{M_w} \ln \varphi$$
(2.13)

where R is the gas constant, M_w is the molar weight of water and ρ is the density of water [2.14].

A relative humidity somewhat less than 100% causes a negative pore water pressure p_w (and partly emptied pores). If that relative humidity is maintained, continuous capillary suction will occur from wetter parts.

In very large pores, the effect of gravity $\rho g h(x)$ must be included, $h(x)$ being the vertical distance between the point x and a water table. In such a case the flow due to capillary suction will be

$$F = -\frac{k_p}{\eta} \frac{d}{dx} [p_w + \rho g h(x)]$$
(2.14)

For an ideal capillary the velocity of the capillary rise is given by [2.15]

$$v = \frac{1}{8\eta} \left(r \times \frac{2\sigma \cos \phi}{x} - g\rho r^2 \right)$$
(2.15)

where
v = velocity of capillary rise (m/s)
η = viscosity of the liquid (Ns/m²)
r = radius of the capillary (m)
σ = surface tension of liquid (N/m)
ϕ = wetting angle
g = gravity (m/s²)
ρ = density (kg/m³)
x = height (m)

Owing to irregularities of the pore system, real materials diverge from this formula. Therefore, empirical relations are established to describe the take-up of a liquid as a function of time.

2.3.1 RATE OF ABSORPTION

For short-term contact of the solid with a liquid, the velocity of the take-up is referred to as initial absorption rate a, given by the mass of liquid absorbed per unit area and a function of the contact time t, for example

$$a = \frac{\Delta m}{Af\left(t^n\right)}$$

(2.16)

where a = absorption rate (g/m^2 s^n)
Δm = take-up of liquid (g)
A = area in contact with water (m^2)
$f(t^n)$ = time function

$\sqrt{(Dt)}$

Frequently, a time function $f(t) = \sqrt{t}$ is valid, then $n = \frac{1}{2}$ [2.16]. However, time functions other than $\sqrt{t}$ may also be chosen if suitable.

2.3.2 NON-STEADY-STATE ABSORPTION

For a short-term contact of a solid surface with water a non-steady-state transport of the liquid prevails: i.e. the amount of liquid absorbed at the surface of the solid as well as the amount of liquid transported at any distance from the surface is a function of time. Under natural exposure conditions this non-steady-state transport occurs if the porous solid is not in permanent contact with the liquid.

2.3.3 STEADY STATE ABSORPTION

Capillary suction may also develop into a steady-state transport phenomenon if suitable boundary conditions are kept constant over time. A concrete member in contact with water on one side will take up the liquid by capillary suction. If evaporation of the water at the opposite side is in equilibrium with the take-up of water, capillary suction transports the water over a certain section of the concrete member in a steady-state process.

2.3.4 MIXED MODES

Capillary absorption is also an important mechanism with respect to the ingress of chlorides into concrete. Non-saturated concrete in contact with a salt solution will take up the salt solution by capillary forces; thus chlorides penetrate into the concrete much faster than by diffusion alone. Simultaneously, chlorides are transported by diffusion to increase the depth of penetration. The simultaneous action of diffusion and capillary suction thus causes a mixed mode of transport, which may be very effective. Also refer to Chapter 6.

2.4 Adsorption and desorption

Adsorption: fixation of molecules on solid surfaces due to mass forces in mono- or multimolecular layers.
Desorption: liberation of adsorbed molecules from solid surfaces.
Major parameters for adsorption and desorption are concentration c of molecules in the gaseous or liquid phase and temperature T [2.17].

2.5 Migration

Transport of ions in electrolytes due to the action of an electrical field as the driving force. In an electrical field positive ions will move preferentially to the negative electrode and negative ions to the positive one.

Migration may generate a difference in concentration in a homogeneous solution or may provoke a species flux in the direction of concentration gradients.

The general law governing mass transfer in electrolytes is given by the **Nernst-Planck equation** [2.18], which has three terms:

$$\text{Flux} = \text{diffusion} + \text{migration} + \text{convection}$$

$$J = D\frac{dC}{dx} + \frac{ZF}{RT}DC\frac{dE}{dx} + CV_e \qquad (2.17)$$

where
J = mass flux (g/m²s)
D = diffusion coefficient (m²/s)
C = concentration (g/m³)
x = distance (m)
Z = electrical charge
F = Faraday constant (J/V.mol)
R = gas constant (J/mol.K)
T = absolute temperature (K)
E = electrical potential (V)
V_e = velocity of solution (m/s)

This general equation states that the movement of charged species in an electrolyte is the sum of the diffusion component from concentration gradients, the migration component arising from potential gradients and the flow of the electrolyte itself (capillary suction, permeation, etc.).

Assuming a stationary flow of ions and provided there is no convection of ions due to electrolyte movement, eq. (2.17) allows the calculation of the diffusion coefficient D.

A direct application of this equation to concrete is difficult because the pore solution is a poly-electrolyte of high ionic strength. However, it can be applied if some simplifications are considered, and therefore very specific tests must be carried out.

2.6   References

2.1.   Lykow, A.W. (1958) *Transporterscheinungen in kapillarporösen Körpern*, Akademie-Verlag, Berlin.
2.2.   Rose, D.A. (1965) Water movement in unsaturated porous materials. *RILEM Bulletin*, No. 29, pp. 119–23.
2.3.   Bazant, Z.P. and Najjar, L.J. (1972) Non-linear water diffusion in non-saturated concrete. *Materials and Structures*, Vol. 5, pp. 1–20.
2.4.   van der Koi, J. (1971) Moisture transport in cellular concrete roofs. Thesis, TH Eindhoven.
2.5.   Crank, J. (1970) *Mathematics of Diffusion*, Oxford University Press.
2.6.   Newman, A.B. (1931) The drying of porous solids. Diffusion calculations. *American Institute of Chemical Engineers*, Vol. 27.
2.7.   Carman, P.C. (1956) *Flow of Gases through Porous Media*, Butterworth Scientific Publications, London.
2.8.   Zagar, L. (1955) Die Grundlagen zur Ermittlung der Gasdurchlässig keit von feuerfesten Baustoffen. *Archiv für das Eisenhüttenwesen*, Vol. 26, No. 12, pp. 777–82.
2.9.   Gertis, K., Kießl, K., Werner, H. and Wolfseher, U. (1976) Hygrische Transportphänomene in Baustoffen, *Deutscher Ausschuß für Stahlbeton*, Heft 258, Verlag Ernst und Sohn, Berlin.
2.10. D'Arcy, H.P.G. (1856) *Les fontaines publiques de la ville de Dijon*, Paris, Dalmont.
2.11. The Concrete Society (1985) Permeability testing of site concrete – a review of methods and experience. Report of a Concrete Society Working Party, in *Permeability of concrete*, Papers of a one-day conference held at Tara Hotel, Kensington, London, 12 December.
2.12. Volkwein, A. (1991) Untersuchungen über das Eindringen von Wasser und Chlorid in Beton, *Berichte aus dem Baustoffinstitut*, Heft 1/1991, Technische Universität München.
2.13. Nilsson, L.-O. Moisture in porous building materials, *Textbook on Building Materials*, Chalmers University of Technology, Göteborg.
2.14. Nilsson, L.-O. (1980) Hygroscopic moisture in concrete - drying, measurement and related material properties, Lund Institute of Technology, Report TVBM-1003, Lund.
2.15. Sommer, E. (1971) Beitrag zur Frage der kapillaren Flüssigkeitsbewegung in porösen Stoffen bei Be- und Entfeuchtungs vorgängen. Dissertation, TH Darmstadt.
2.16. Hall, C. (1989) Water sorptivity of mortars and concretes, a review. *Magazine of Concrete Research*, Vol. 41, No. 147, pp. 51–61.

2.17. Brunauer, S., Emmet, P.H. and Teller, E. (1938) Adsorption of gases in multimolecular layers. *J. American Chemical Society* Vol. 60, No. 2, pp. 309–19.

2.18. Bockris, J.O.'M. and Reddy, A.K.N. (1974) *Modern Electrochemistry*, Plenum Press, New York.

3 Relations between different transport parameters

Lars-Olof Nilsson and Tang Luping

The flow of liquids, gases and ions in a specific concrete should be inter-related to some extent, since these flow processes occur in the same pore system, usually in different parts of it, however. In this chapter some theoretical relationships are shown and available data are presented.

3.1 Permeability vs diffusion coefficient in general

The general relationship between permeabilities K (m^2), K_w (m^2/s) and diffusion coefficients D (m^2/s) may be derived as follows.

Provided that Hagen-Poiseuille's law is valid also in small pores, the permeability K of a single straight pore with radius r_{eff} embedded in a medium of cross-sectional area A can be simply expressed as

$$K = \frac{\pi r_{eff}^4}{8A} \tag{3.1}$$

The diffusion coefficient D can be described as

$$D = D_0 a_{eff} = D_0 \frac{\pi r_{eff}^2}{A} \tag{3.2}$$

where a_{eff} is the area fraction of effective pores and D_0 is the diffusion coefficient in a bulk fluid.

If the effective radius r_{eff} is the same in eqs (3.1) and (3.2), then

$$K = \frac{r_{eff}^2}{8D_0}D = \frac{r_{eff}^2}{constant}D \tag{3.3a}$$

$$K = \frac{A}{8\pi D_0^2}D^2 = constant \times D^2 = constant \times D^b \tag{3.3b}$$

Performance Criteria for Concrete Durability, edited by J. Kropp and H.K. Hilsdorf. Published in 1995 by E & FN Spon, London. ISBN 0 419 19880 6.

where $b = 2$ for cylindrical pores with similar r_{eff} for both permeation and diffusion. Equation (3.3a) corresponds to the Katz-Thompson equation [3.1]. Equation (3.3b) corresponds to the findings of Gaber [3.2].

Different flow mechanisms, cracks, different fluids and different substances being transported may cause different r_{eff}. This could be looked upon as being described with different b values in eq. (3.3b).

3.2 Moisture and water flow

Water may move in pore systems due to differences in pore water pressures and water vapour pressures or contents as saturated water flow, capillary suction, pore surface flow and vapour diffusion. The processes occur more or less simultaneously in a series and parallel pore system. The division between them is difficult and they cannot be measured separately. The total flow of water is measured, and the conditions set up for such a measurement may give a main part of the total flow as one or two of these processes.

The **water vapour diffusion coefficient** D_c describes a pure gas diffusion process in an almost completely empty pore system (see Fig. 3.1) *if* it is measured at very low moisture conditions.

The low moisture condition, however, may add cracks to the pore system that are not present in the same system at higher humidities.

$$K = \text{constant} \times D_c^b \tag{3.4}$$

3.2.1 WATER PERMEABILITY VS VAPOUR DIFFUSION

The **saturated water flow permeability** K describes a hydraulic flow in the same pore system as the diffusion of water vapour, excluding drying cracks. The relative humidity in this case is RH = 100% (Fig. 3.2).

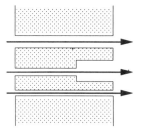

Fig. 3.1. Vapour diffusion in empty pore system.

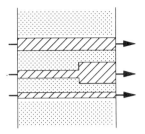

Fig. 3.2. Water flow in saturated pore system.

A relationship between the permeability and the diffusion coefficient according to eq. (3.4) should be expected, provided that the diffusion co-efficient is measured at very low humidities.

Some measured data for this relationship are shown in Fig. 3.3. In this case a relation according to eq. (3.4) exists with $b \approx 1.8$.

3.2.2 MOISTURE FLOW COEFFICIENT VS WATER PERMEABILITY

The **moisture flow coefficient** D_c at an intermediate RH is not uniquely related to water flow and vapour flow permeabilities if $50\% < RH < 100\%$. The flow mechanisms are different in different pores, depending on the pore size and the moisture distributions (see Fig. 3.4).

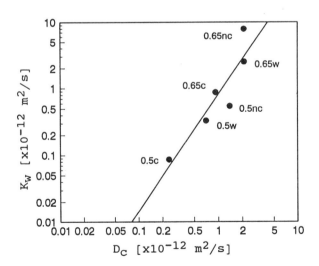

Fig. 3.3. Measured relation between water permeability K_w and water vapour diffusion coefficient D_c (RH = 0.39) of cement mortars [3.3].

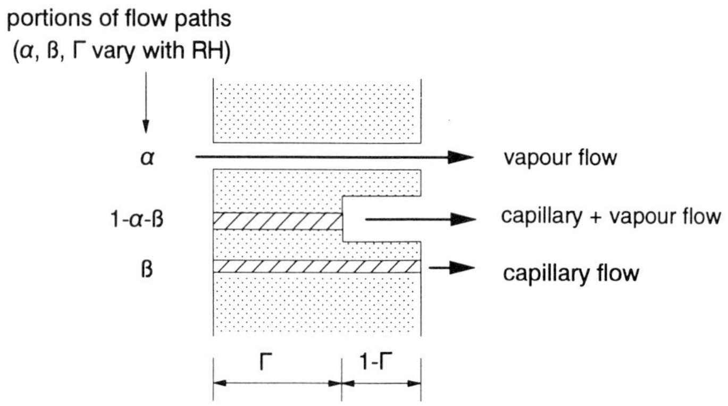

portions of flow paths
(α, β, Γ vary with RH)

Fig. 3.4. Moisture flow in pore system at an intermediate RH.

In Fig. 3.5 measured permeabilities and moisture flow coefficients at RH > 90% are shown. As expected, there is no unique relationship. At 100% RH the moisture flow coefficient D_c, by definition, should be proportional to the hydraulic permeability K at a pore water pressure of zero. The relation may be found from

$$q = D_c \frac{\Delta c}{\Delta x} = \frac{K}{\eta} \rho \frac{\Delta p_w}{\Delta x} \qquad (\mathrm{kg}/\mathrm{m}^2 \mathrm{s})$$

(3.5)

which gives, at RH = 100%,

$$K_{(p_w=0)} = D_{c(\mathrm{RH}=100\%)} \frac{\eta}{\rho} \frac{M_w c_s(T)}{\rho RT}$$

(3.6)

No data on this relationship have been found in the literature. The obvious reason for this is that $D_c(\mathrm{RH} = 100\%)$ is very difficult to measure directly.

3.2.3 RATE OF WATER ABSORPTION VS WATER PERMEABILITY

The take-up of liquid, Δm, during contact with water during a period of time t is given by

$$\Delta m = \int_0^t FA\,dt$$

(3.7)

or

$$\frac{\Delta m}{A} = \int_0^t F\,dt = a\,f(t)$$

(3.8)

cf. eq. (2.16), where a is the rate of water absorption.

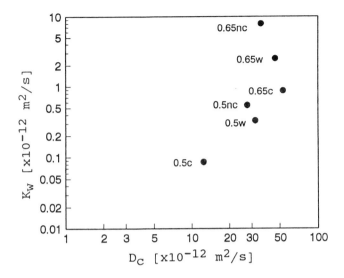

Fig. 3.5. Measured permeabilities K_w and moisture flow coefficients D_c at RHs greater than 90% [3.3].

The flow $F(t)$ decreases with time. To predict the flow, Fick's second law has to be solved with a proper flow description and relevant initial and boundary conditions.

With a flow description in terms of water content w as the concentration and D_w the 'diffusion coefficient', one example of a solution is shown in Fig. 3.6.

The assumption is that

$$D_w = D_c^{TOT} \cdot \frac{c_s}{dw/d\varphi} = k_p^{TOT} \frac{1}{\partial w/\partial p_w} = \text{constant} \tag{3.9}$$

where $dw/d\varphi$ and dw/dp_w are the slopes of the sorption isotherm $w(\varphi)$ and the suction curve $w(p_w)$, respectively. These slopes are not generally constant but vary significantly with the moisture content.

The integral in eq. (3.8) can be rewritten, if the moisture content w_0 is evenly distributed before the uptake:

$$\int_0^t F dt = LU_m \left(w_{cap} - w_0 \right) \tag{3.10}$$

The relation between the rate of absorption a and the capillary liquid flow permeability is consequently

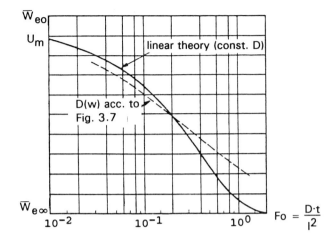

Fig. 3.6. A solution to Fick's second law for a slab with thickness L and sealed upper surface.

$$a = \frac{LU_m(F_0)(w_{cap} - w_0)}{f(t)}$$

(3.11)

where U_m is the solution to Fick's second law
      F_0 is the Fourier number $D_0 t/L^2$; $D_0 = D_w(w_0)$
      D_w is the diffusion coefficient in eq. (3.9) and a function of the permeability
      w_{cap} is the moisture content at capillary saturation
      w_0 is the original moisture content before the water uptake
      $f(t)$ is the time function (usually t) in the definition of a; cf. eq. 2.16

If, additionally, $U_m(D_w, t)$ and $f(t)$ follow the same time function, the rate of absorption a would be a constant. The time function $f(t)$ is usually **chosen** such that a will be constant. Usually, U_m is proportional to $\sqrt{t}$ provided the original moisture content was evenly distributed and as long as the moisture has not reached the upper surface. Then a will be a constant if $f(t) = \sqrt{t}$.

Consequently the relation between the rate of absorption a and the water permeability k depends 'only' on:

1. the slope and shape of the suction curve, which gives the 'moisture capacity' dw/dp_w, w_{cap} and w_0;
2. The chosen time function $f(t)$;
3. The moisture dependence of the coefficient of moisture permeability.

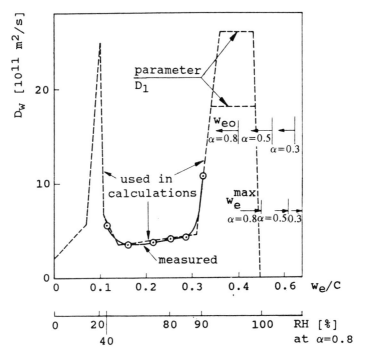

Fig. 3.7. An example of non-constant $D_w(w)$ [3.4].

$$k_p^{TOT} = k_p(w_{cap}) \left[\frac{k_p(w)}{k_p(w_{cap})} \right]$$

(3.12)

With the **assumption** that the permeability and the slope of the suction curve vary in the same way with the moisture content, D_w would be a constant. If not, another solution $U_m(t)$ must be found; and maybe another $f(t)$!

For concrete D_w is not a constant. One example of the $D_w(w)$ is shown in Fig. 3.7. This would give another $U_m(t)$ but with a proper choice of $f(t)$, a would still be a constant.

A close relationship between the water permeability K ($w = w_{cap}$), measured at low pressures, and moisture permeability at saturation is expected, and would be possible to find from eq. (3.9), eq. (3.12), and the solution $U_m(t)$ to Fick's second law for a proper $D_w(w)$. This is, however, not easy and has so far never been done.

3.3 Gas flow

3.3.1 GAS PERMEABILITY VS GAS DIFFUSION

The 'permeability' of a gas is a measure of the flow at an abnormally high gas pressure. The 'diffusion coefficient' of a gas is measured at normal partial pressures of the gas. They should be related to each other since the flow processes occur in the same pore system provided drying is prevented during the measurement or both measurements are made at dry conditions.

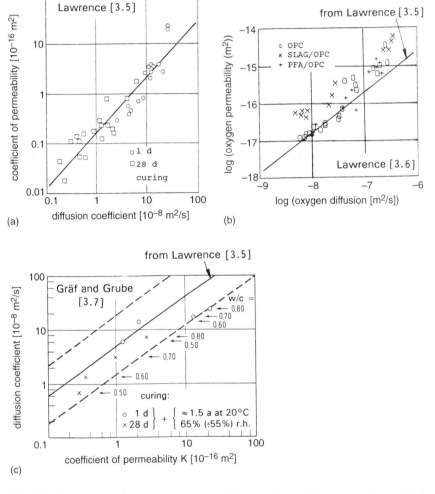

Fig. 3.8. The relation between gas permeability and gas diffusion coefficient [3.5, 3.6, 3.7].

The relationship that can be expected should be according to eq. (3.3b),

$$K_g = \text{constant} \times D^b$$

(3.13)

Some experimental data are shown in Fig. 3.8. For the data in Fig. 3.8, $b \approx 1$.

3.3.2 EFFECT OF MOISTURE CONTENT

The flow of gases is of course much more rapid in empty pores than in pores filled with water; see Fig. 3.9.

Dissolved gases may also diffuse through the liquid phase; however, the rate of diffusion is reduced then by a factor of 10^5 as compared with the diffusion in the gaseous phase.

At different moisture contents the gas flow coefficient gives a rough estimate of continuous 'air paths' including cracks (see Fig. 3.10). This should be related to the water vapour diffusion at low humidity conditions and the hydraulic permeability at water pressures larger than zero when both flows occur also in cracks; see section 3.3.4 below.

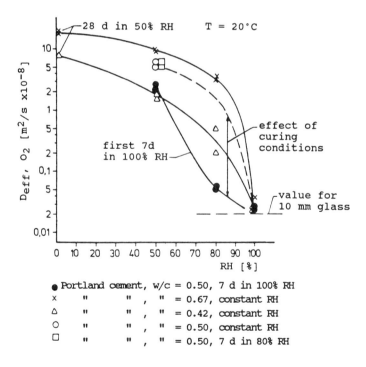

Fig. 3.9. The effect of moisture on the diffusion of oxygen [3.8].

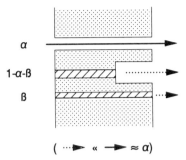

Fig. 3.10. Gas flow in pore system at an intermediate RH.

3.3.3 DIFFUSION OF DIFFERENT GASES

The diffusion coefficients for different gases might be different, owing to the size of the molecules. According to the kinetic theory of gases, the effective diffusivity is inversely proportional to the square root of the molecular weight M.

An example is shown for the diffusion of CO_2 and of O_2 through aerated concrete in Fig. 3.11. The measurements in Fig 3.11 seem to coincide with the effect of molecular weight on the effective diffusivity up to some 60% RH. At higher humidities, however, some other effect probably influences the dissolution of CO_2 in the pore water.

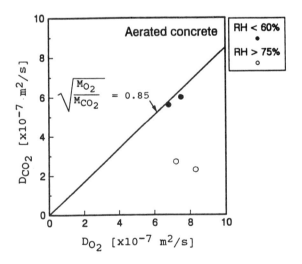

Fig. 3.11. Diffusion coefficients for CO_2 and O_2 [3.9].

3.3.4 GAS PERMEABILITY VS WATER PERMEABILITY

The permeation of gases and liquids is described in the same way, with permeability coefficients K_g and K respectively. In the same pore system these should be equal: i.e. if the gas permeability is measured in a dry pore system and no drying cracks are added, then

$$K_g(\text{RH} = 0) \approx K \qquad (3.14)$$

A certain moisture content in the pore system should change eq. (3.14) into

$$K_g(\text{RH}) < K \qquad (3.15)$$

The flow of gas and the flow of liquid have different resistances from the pore wall surfaces [3.10]. Since the concentration of molecules is much larger in a liquid, the resistance to liquid flow is higher. This resistance should have largest effect in smaller pores, i.e.

$K_g \approx K$ for large radii
$K_g > K$ for small radii

This latter relation is clearly shown in the data presented in Fig. 3.12.

3.3.5 GAS PERMEABILITY VS RATE OF WATER ABSORPTION

Since there is a certain relationship between the gas permeability and water permeability, cf. section 3.3.4, and between water permeability and rate of water absorption, cf. section 3.2.3, we would expect a relationship between gas permeability and rate of water absorption. An experimental relationship does exist; see Fig. 3.13.

3.4 Ion diffusion

Ions move in a continuous water path (see Fig. 3.14) and, to some extent, in the adsorbed layer at pore surfaces. They move

- with the water flow,
- as diffusion without a water flow.

In this section only diffusion is dealt with.

The ion diffusion coefficient is an estimate of the continuous water paths.

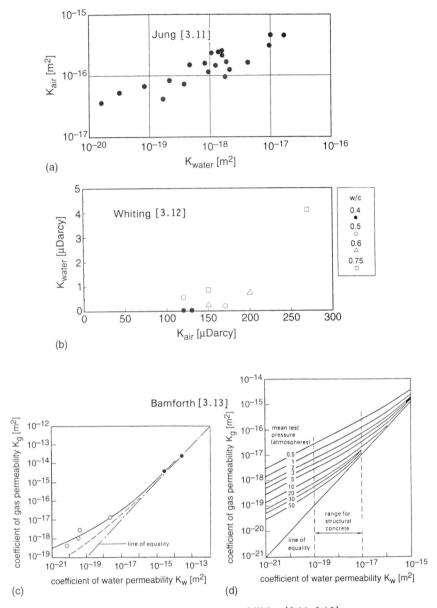

Fig. 3.12. Gas permeabilities vs water permeabilities [3.11–3.13].

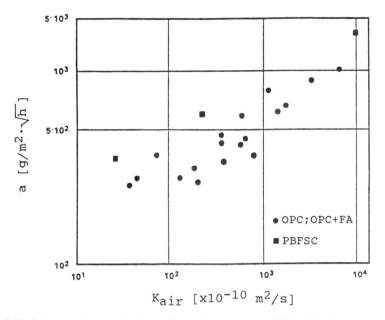

Fig. 3.13. Measured rates of absorption and gas permeabilities [3.14].

3.4.1   ION DIFFUSION VS WATER PERMEABILITY

At saturated conditions the steady-state flow coefficient should be related to the water permeability since the processes occur in the same system. The relation should be as in eq. (3.3b):

$$K = \text{constant} \times D_{ion}^{b} \tag{3.16}$$

Two examples are shown in Fig. 3.15.

The relations in Fig. 3.15a give a value of $b \approx 1.5$ in eq. (3.16). This is, however, an example for sand and for Ca^{2+} ions. Similar measurements

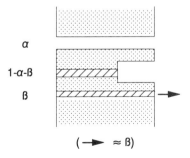

Fig. 3.14. Ion flow in continuous water paths.

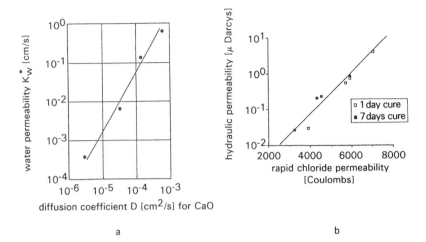

Fig. 3.15. Relations between water permeability and ion diffusion coefficient: (a) by Grube and Rechenberg [3.15], and (b) by Whiting [3.12].

for concrete have not been found. Narrow pores should, however, affect the diffusion of ions and the liquid flow differently. This means that for dense concretes the relation should be different from that in Fig. 3.15a. In addition, the value of b in eq. (3.16) depends on the type of ions since the ion mobility is different for different ions.

3.4.2 DIFFERENT IONS

The diffusion coefficients for different ions in bulk water are different owing to different ionic mobilities and ion valences. From Atkins [3.16] the data in Table 3.1 are derived. Diffusion coefficients for different ions in concrete should also be affected by the mobility of the ion and the presence of other ions. An example is shown in Fig. 3.16.

Since the penetration depth, at a certain time, is proportional to the square root of the diffusion coefficient, we would expect a relationship between the penetration depths of different ions. An example is shown in Fig 3.17.

In addition, the ion binding capacity of concrete, the type and concentra-

Table 3.1. Diffusion coefficients D_0 in bulk water at 25°C

ION	Cl^-	Na^+	SO_4^{2-}
D_0 ($\times 10^{-9}$ m²/s)	2.03	1.33	1.06
Radius (Å)	1.81	0.95	–

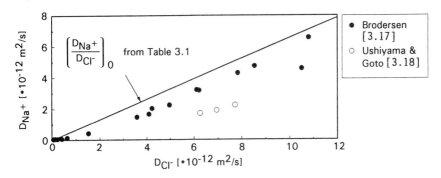

Fig. 3.16. An example of the relation between diffusion coefficients for different ions in concrete.

tion of ions in the pore solution, or the interaction between ions and pore surface charges, could also affect the ion diffusion. Since the binding layer can block the continuity of the pore system and the surface charges can retard the diffusion of a charged ion, it could be expected that the influence of binding and surface charges on diffusion coefficient in denser concrete is more significant, as shown in Fig. 3.18. The oxygen diffusion in Fig. 3.18 occurs in a saturated pore system.

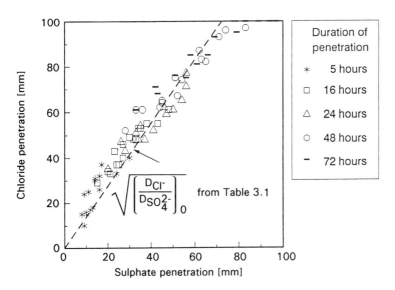

Fig. 3.17. The penetration of different ions in concretes with $w/c = 0.4 \sim 0.62$; [3.19, 3.20].

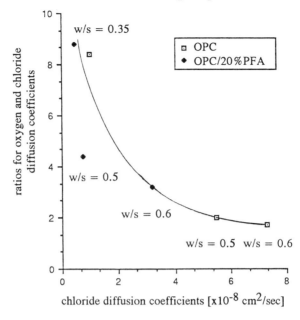

Fig. 3.18. Influence of ion binding and surface charges on diffusion coefficient in saturated pastes [3.21].

3.5 Conclusion

Generally the theoretical relationships between different flow coefficients are scarcely studied and experimental data are lacking. However, the relation between permeabilities K and diffusion coefficients D may be expressed in general as

$$K = \text{constant} \times D^b$$

where the parameter b depends on the transported substances in the comparison. Limited experimental data show values of b according to Table 3.2.

Table 3.2. Experimental b values

Substance in permeation	diffusion	b
Water	Water vapour	1.8
Gas	Gas	1.0
Water	Ions	1.5

'Indirect' flow coefficients, such as the rate of water absorption *a*, may have an empirical relationship to other flow coefficients, but the theoretical relationship is complicated and depends on a number of material properties.

The gas flow coefficients are highly dependent on the degree of blocking of the pore system by moisture. Consequently, the relationship between these flow coefficients and others depends very much on the moisture content.

Physical and chemical interaction between the diffusing or permeating species and the pore surfaces may influence the flow of different gases and different ions in different ways. In addition, such an interaction may cause variations in the relationships with pore size distribution and chemical composition of the binder.

3.6 References

3.1. Garboczi, E.J. (1990) Permeability, diffusivity, and microstructural parameters: A critical review. *Cement and Concrete Research,* Vol. 20, No. 4, pp. 591–601.

3.2. Gaber, K. (1988) Influence of mix proportions and components on the diffusion coefficient and the permeability of concrete. *Darmstadt Concrete*, 3, pp. 39–48.

3.3. Kropp, J. (1983) Karbonatisierung und Transportvorgänge in Zementstein. Dissertation Universität Karlsruhe.

3.4. Nilsson, L.-O. (1980) *Hygroscopic moisture in concrete – drying, measurements and related material properties*, TVBM-1003, Lund Institute of Technology.

3.5. Lawrence, C.D. (1984) Transport of oxygen through concrete. *British Ceramics Society Proceedings*, No. 35, pp. 277–93.

3.6. Lawrence, C.D. (1986) Measurement of permeability, in *Proc. 8th Intl. Congress on Chemistry of Cement*, Brazil, Vol. I, pp. 29–34.

3.7. Gräf, H. and Grube, H. (1986) Einfluß der Zusammensetzung und der Nachbehandlung des Betons auf seine Gasdurchlässigkeit. *Beton*, Vol. 36, No. 11, pp. 426 429 and No. 12, pp. 473–6.

3.8. Tuutti, K. (1982) *Corrosion of Steel in Concrete*, Swedish Cement and Concrete Research Institute (CBI), Stockholm, No.4.82.

3.9. Houst, Y. and Wittmann, F.H. (1986) The diffusion of carbon dioxide and oxygen in aerated concrete, in *Int. Coll. on Materials Science and Restoration*, Technische Akademie, Esslingen.

3.10. Tang, L. and Nilsson, L.-O. (1992) A study on the quantitative relationship between permeability and pore size distribution of hardened cement pastes. *Cement and Concrete Research*. Vol. 22, No. 4, pp. 541–50.

3.11. Jung, M. (1969) Beiträge zur Gütebewertung korrosions- und wasserdichter Betone. Dissertation Weimar.

3.12. Whiting, D. (1988) Permeability of selected concretes, in *Permeability of*

Concrete, D. Whiting and A. Walitt (Eds), ACI Special Publication, SP-108, Detroit, pp. 195–222.

3.13. Bamforth, P.B. (1987) The relationship between permeability coefficients for concrete obtained using liquid and gas. *Magazine of Concrete Research*, Vol. 39, No. 138, pp. 3–11.

3.14. Hilsdorf, H.K., Schönlin, K. and Burieke, F. (1991) *Dauerhaftigkeit von Betonen*, Universität Karlsruhe.

3.15. Grube, H. and Rechenberg, W. (1987) Betonabtrag durch chemisch angreifende saure Wässer. *Beton*, Vol. 37, No. 11, pp. 446–51 and No. 12, pp. 495–8.

3.16. Atkins, P.W. (1978) *Physical Chemistry*, Oxford University Press.

3.17. Brodersen, A. (1982) Zur Abhängigkeit der Transportvor-gänge verschiedener Ionen im Beton von Struktur und Zusammensetzung des Zementsteins. Dissertation, RWTH Aachen

3.18. Ushiyama, H. and Goto, S. (1974) Diffusion of various ions in hardened Portland cement paste, in *Proc. 6th Intl. Congress on Chemistry of Cement*, Moscow, Vol.II-1, pp. 331–7.

3.19. Rio, A. and Turriziani, R. (1981) The influence of superplasticizing agents on the penetration of aggressive ions into cement concretes and on their resistance to attack by aggressive solutions. *Il Cemento*, Vol. 4, pp. 171–82.

3.20. Rio, A. and Turriziani, R. (1982) New studies on the influence of the reduction of the water/cement ratio on the chemical resistance of cement concretes. *Il Cemento*, Vol. 1, pp. 45–58.

3.21. Yu S.W. and Page, C.L. (1991) Diffusion in cementitious materials: 1. Comparative study of chloride and oxygen diffusion in hydrated cement pastes. *Cement and Concrete Research*, Vol. 21, No. 4, pp. 581–8.

4 Parameters influencing transport characteristics

Jean-Pierre Ollivier, Myriam Massat and Leslie Parrott

4.1 Technological parameters (materials and execution) influencing water and gas permeability

Concrete permeability depends on a vast range of parameters, which can be grouped together into several categories:

- constituents characteristics;
- sample characteristics;
- stress conditions.

The diagram in Fig. 4.1 is based on that presented by Mehta [4.1] concerning the mechanical strength of concrete. The main features of permeability shown here have a role to play in concrete porosity (capillary voids or

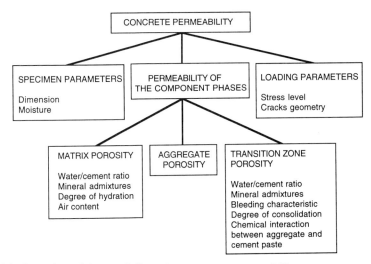

Fig. 4.1. Interplay of factors influencing concrete permeability.

Performance Criteria for Concrete Durability, edited by J. Kropp and H.K. Hilsdorf. Published in 1995 by E & FN Spon, London. ISBN 0 419 19880 6.

microcracks) and in possible fluid transfer across the material. It is for these reasons that they influence permeability. This report deals solely with the influence of the type and proportions of the constituents, as well as that of age and curing conditions.

4.1.1 PRELIMINARY REMARKS: PURE PASTE AND CONCRETE PERMEABILITY

The studies analysed for the purposes of this report have been carried out either on pure cement paste or on concrete. In the latter case, permeability depends on the nature, quantity and size of aggregates in addition to the characteristics of the binding phase.

Theoretically, adding low-permeability aggregates to cement paste should reduce overall permeability by interrupting capillary pore continuity in the cement paste matrix, particularly in the case of young paste with a high water/cement ratio, and high capillary porosity. In consequence, concrete and mortar of the same age and water content should be less permeable than pure paste. Results of tests, however, indicate that the opposite is true. Data in Fig. 4.2 show the considerable increase in permeability when aggregates are added to a paste or mortar [4.2, 4.3]. The greater the size of the aggregates, the greater is the permeability. These results have been analysed by Mehta [4.1], who attributes the permeability increase to the presence of microcracks in the transitional zone at the interface between paste and aggregates [4.4]. In addition, two phenomena contribute to this permeability increase:

- the interface microcracking becomes more severe as the maximum aggregate size increases;
- the porosity and the pore size are greater in this transition zone than in the bulk.

Dhir *et al.* [4.5] found no significant difference in air permeabilities of concrete made from aggregates of differing sizes (up to 20 mm). The permeability increase observed when using 40 mm aggregates is considered to be due to the lower quality of the paste–aggregate interface.

These preliminary remarks thus show that, leaving aside metrological problems, it will often be difficult to compare results from different studies. It is also difficult to analyse the effect of one particular composition parameter from literature data in so far as several parameters are usually varying simultaneously. This is probably the case for the study reported in [4.2], in which the cement content is certainly varying with the maximum aggregate size.

The analysis of aggregate effect on permeability may be completed by Nyame's results [4.6]. The permeability of normal and lightweight mortars (prepared at a constant water/cement ratio) is varying with the aggregate

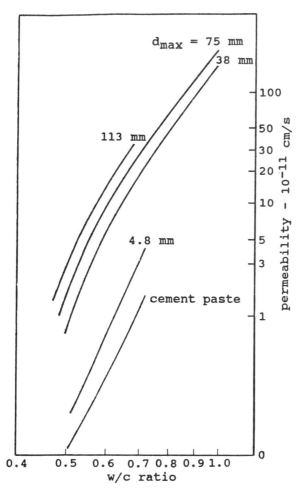

Fig. 4.2. Influence of water/cement ratio and maximum aggregate size upon the water permeability of concrete [4.2].

volume concentration as shown in Fig. 4.3. Nyame concludes as follows:

- At increasing aggregate volume concentrations, interfacial effects increase the permeability, whilst absorption of paste water by the aggregate reduces the permeability of mortars.
- The permeability of mortars increases as porosity reduces, contrary to the response of hardened cement pastes (an opposite result has been presented by Watson *et al.* [4.7]).
- Lightweight mortar is about twice as permeable as, and not of a different order of permeability from, sand mortar at a given aggregate volume concentration.

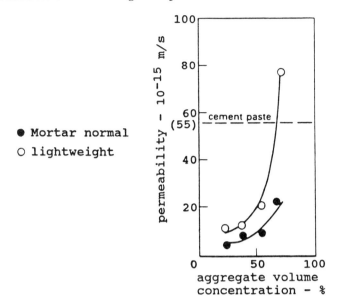

● **Mortar normal**
○ **lightweight**

Fig. 4.3. Effect of aggregate volume concentration upon the water permeability of mortars [4.6].

4.1.2 PERMEABILITY OF PORTLAND CEMENT CONCRETE WITHOUT ADDITIONS

Fluid transfers across concrete are fundamentally the same as for all porous materials. Powers [4.8] has shown a relationship between permeability and capillary porosity (Fig. 4.4). Gel pores also contribute to the possibilities of transfer across cement paste, but in a very limited way. According to Powers [4.8] the water permeability of this phase is only 7×10^{-16} m/s. Mehta [4.1] has likewise shown the influence of cement paste hydration and of the water/cement ratio on capillary porosity. Permeability of hardened cement paste (hcp) is practically nil for capillary porosities of less than 10%, which corresponds to overall porosity in the region of 35% [4.9]. This threshold is generally interpreted as a transitional stage between a system of interconnected pores and a system of isolated pores. With equal overall porosity, permeability depends on pore size distribution, pore continuity and isotropy [4.10]. According to Massazza [4.11], concrete permeability depends largely on pores whose diameter exceeds 0.1 μm; an essential role is also played by their shape and connectivity.

Mention is also made of the influence of the specific area of hydration products [4.12]; cement pastes with a large specific surface area can trap a large amount of water, thus slowing down the migration of humidity in pastes with a high capillary porosity [4.13].

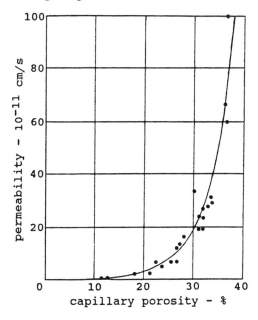

Fig. 4.4. Relationship between water permeability and capillary porosity for Portland cement pastes [4.8].

4.1.3 FACTORS INFLUENCING PERMEABILITY OF PORTLAND CEMENT CONCRETE

(a) Water/cement ratio

For pastes with the same degree of hydration, permeability is reduced as the water/cement ratio decreases. The results of Powers *et al.* [4.14] confirm this for pastes with a hydration level of 93% (Fig. 4.5).

Results for concrete are not so easy to analyse, owing to the wide range of composition; however, the results reported by Lawrence [4.15] confirm the trend for the influence of the water/cement ratio: the water permeability increases with the water/cement ratio. Nevertheless, the influence of the water/cement ratio in itself is not clearly quantified in the available data because other parameters are also varying (cement content, aggregate grading).

(b) Age

As capillary porosity decreases with the progression of hydration reactions, so paste and concrete permeability decrease with age. Gräf and Grube [4.16] reproduce this schematically in Fig. 4.6.

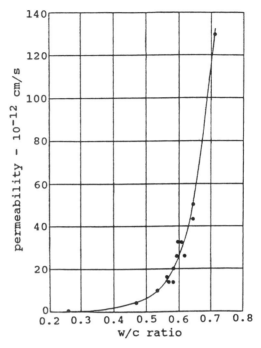

Fig. 4.5. Relationship between water permeability and water/cement ratio for mature cement pastes [4.14].

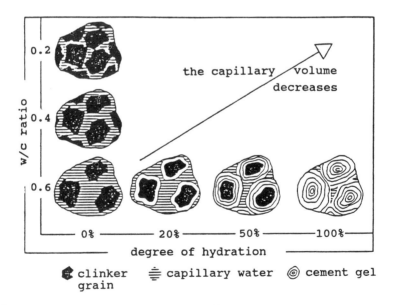

Fig. 4.6. Modelling of capillary filling [4.16].

Table 4.1. Reduction in permeability of cement paste (w/c = 7.0) with the progress of hydration

Age (days)	Permeability (10^{13} m/s)
Fresh	20 000 000
5	4 000
6	1 000
8	400
13	50
14	10
Ultimate	6

Table 4.1 shows the typical results of the influence of the state of hydration on pure paste permeability (w/c = 0.7).

The permeability coefficient of pastes immediately after mixing is in the region of 10^{-5}–10^{-6} m/s. As with capillary porosity, permeability decreases over a period of time, but the two are not directly proportional to each other. In fact, at the start of hydration, there is a slight decrease in capillary porosity together with a significant segmentation of the large pores, thus considerably reducing the possibilities of fluid transfer across the cement paste. Mehta [4.1] indicates that beyond a threshold of 30% capillary porosity, the pores are so tortuously interconnected that a decrease in paste porosity does not result in a significant decrease in permeability. Under these conditions, permeability is in the region of 10^{-14} m/s. The above threshold is reached after 3, 14, 180 or 365 days in pastes with a water/cement ratio of 0.4, 0.5, 0.6 or 0.7 respectively.

Uchikawa [4.17] explains that the decrease in cement paste permeability with hydration bears no direct relation to the amount of gel formed. It depends rather on where the gel formation is sited. When these sites result in capillary blockage, permeability is reduced. Portlandite does not affect capillary continuity in the same way. This can be compared with the classic analysis of Powers *et al.*, which shows that an increase in water/cement ratio leads to an increase in the time taken for capillaries to segment [4.18]: see Table 4.2.

Table 4.2. Segmentation time of the capillary network

w/c radio	Time
0.40	3 days
0.45	7 days
0.50	14 days
0.60	6 months
0.70	1 year
> 0.70	Impossible

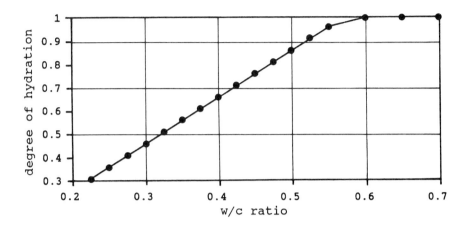

Fig. 4.7. Degree of hydration required to achieve capillary pore discontinuity for various *w/c* ratios [4.19].

The information given in Tables 4.1 and 4.2 may be presented in terms of hydration required to achieve pore discontinuity. A calculation of the degree of hydration required to achieve capillary pore discontinuity for cement paste at any given *w/c* ratio has been presented by Bentz and Garboczi [4.19]. The curve presented in Fig. 4.7 is plotted from the results of this calculation.

The above results relate to a particular type of Portland cement, and notably to its reactivity. Cement evolved in such a way that, in order to obtain a specified strength resistance, increasing water/cement ratios are possible. This can lead to an increase in permeability. Parrott [4.20] has analysed cement evolution data obtained between 1950 and 1980 [4.21]. Assuming that the size distribution of cement particles remains unchanged during the same period, he calculated, for specified compressive concrete strength, the moist curing time in such a way as to obtain low permeability at a given value (chosen arbitrarily at 28×10^{-15} m/s). The results are given in Table 4.3.

Water permeability of cement pastes between 2 days and 10 months has been measured by Nyame and Illston [4.22]. Their results show (Fig. 4.8) that the water/cement ratio (*w/c* ratio from 0.23 to 1.00) does not significantly influence the rate of decrease of permeability over a

Table 4.3. Curing time needed to reduce permeability to
28×10^{-15} m/s

28-day cube strength (MPa)	1950 OPC		1980 OPC	
	w/c	days	w/c	days
25	0.62	80	0.73	900
35	0.51	16	0.59	18
45	0.43	5.7	0.50	4.5
55	0.38	3.3	0.43	2.0

period of time. These results contradict Mehta's analysis, particularly in the case of the permeability decrease observed after 28 days with a water content level of 0.23. Schönlin and Hilsdorf's results [4.23] agree more closely with Mehta: permeability expressed by means of an index stabilizes much more rapidly in a concrete with a water/cement ratio of 0.45 than in one with a water/cement ratio of 0.60.

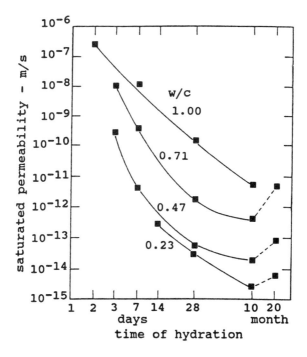

Fig. 4.8. Effects of hydration on water permeability of hardened cement pastes of different *w/c* ratios [4.22].

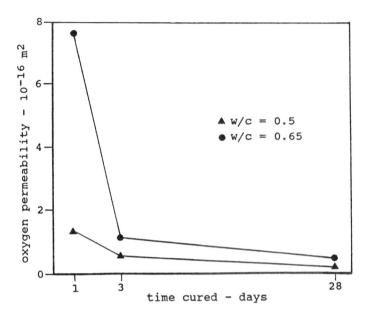

Fig. 4.9. Oxygen permeability of concretes as a function of curing time and *w/c* ratio [4.24].

Grube and Lawrence [4.24] have published the results of an interlaboratory analysis in which oxygen permeability for concrete values are contained within the range: 0.001 and 30×10^{-16} m². The report provides an average permeability reading obtained in different laboratories for two concrete mixtures (with a *w/c* ratio of 0.5 or 0.65, and with a cement paste volume of 295 l/m³) according to the curing time. The results (Fig. 4.9) agree with those of Schönlin and Hilsdorf and show that reducing water content, at a given volume of paste per m³ of concrete, leads to a material that is less sensitive to an initially poor curing. It is concluded that since oxygen permeability measurements are sensitive to slight variations in composition and curing, they can serve as a good indicator of concrete quality.

(c) Curing

The effect of an initial moist curing for concretes prepared with different *w/c* ratios has been studied by Dhir *et al.* [4.5]. Air permeability coefficients measured at 28 days increase almost exponentially with the *w/c* ratio. The time of curing also plays an important part, as shown in Fig. 4.10. 28-day air permeability increases as the time of initial curing in water diminishes: the greater the water content, the greater the in-

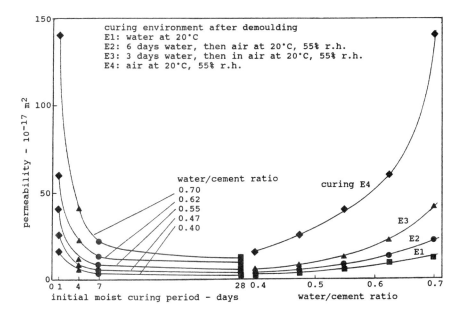

Fig. 4.10. Effect of initial moist curing period on air permeability [4.5].

crease. The authors also mention that if the compressive strength of a particular concrete is closely related to its water content and the curing conditions, it is not however possible to predict permeability from a given strength.

Marsh [4.25] has measured water permeability of pure pastes with *w/c* of 0.47 (curing having taken place at 20, 35, 50 and 65°C). 28-day and 3-month-old concretes kept at 20°C have a mean permeability of approximately 1.5×10^{-13} m/s. With curing at 35, 50 and 60°C, water permeabilities vary between 4 and 9×10^{-12} m/s.

Other tests on pastes with *w/c* of 0.35 give at 28 days a water permeability of 6×10^{-14} m/s at 27°C and 10^{-12} m/s at 60°C [4.26].

The action of elevated temperatures on hardened concrete may result in micro-cracking, which in turn increases permeability. Häkkinen mentions an air permeability increase factor of 5 for concrete treated at 10 days for a period of 6 days at 40°C [4.27].

(d) Aggregate type

The porosity of ordinary aggregates is much less than the capillary porosity of cement paste. However, their permeability is not as different as the

Table 4.4. Comparison of the permeability of different types of rock and cement paste

Type of rock	permeability (cm/s)	w/c ratio of mature paste of the same permeability
Dense trap	2.47×10^{-12}	0.38
Quartz diorite	8.24×10^{-12}	0.42
Marble	2.39×10^{-11}	0.48
Marble	5.77×10^{-10}	0.66
Granite	5.35×10^{-9}	0.70
Sandstone	1.23×10^{-8}	0.71
Granite	1.58×10^{-8}	0.71

porosity discrepancies would lead one to suppose (Table 4.4). These results are explained by a different distribution of pore size: in the region of 10–100 nm for paste capillaries and generally in excess of 10 μm for aggregate pores.

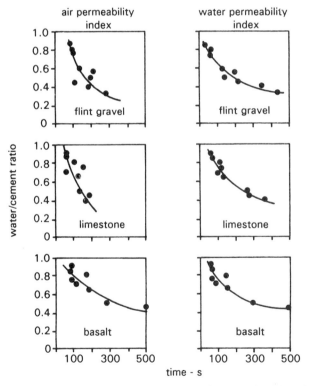

Fig. 4.11. Air and water permeability of concrete (index using the Figg method) as a function of w/c and type of aggregate [4.28].

Few studies have been made on the influence of aggregate type. The only results are from on-site measurements obtained a long time ago by Figg [4.28]. The water permeability rating is hardly affected, whereas air permeability decreases with the use of calcareous aggregates (Fig. 4.11).

(e) Entrained air

Kasai [29] has shown that on-site permeability was little affected by the addition of entrained air in high slump (20 cm) concrete whereas the same addition in 8 cm slump concrete leads to a permeability decrease due to the densification process which accompanies the air entraining effect.

(f) Additions

Cement paste containing additions has a lower water permeability than Portland cement pastes [4.30, 4.31, 4.32]. This reduction can be of three orders of magnitude [4.33]. This reduction of permeability is often attributed to additions (pozzolana, fly ash or slag), which enhance the formation of smaller and less permeable capillaries [4.30]. Alternatively, it is suggested that, on account of the greater porosity and larger pore size, pore continuity is reduced [4.32, 4.34].

This reduction of water permeability is also found in concrete [4.35]. Oxygen permeability measurements indicate that in fact this conclusion depends on concrete maturity. Massazza [4.36] points out that with a sufficient quantity of cement, and a prolonged moist curing, the difference between Portland cement and blended cement disappears. A longer curing is necessary because pozzolana and fly ash react more slowly than Portland clinker. This analysis is based on Gräf and Grube's results [4.37] summarized in Tables 4.5 and 4.6 and in Fig. 4.12 [4.36].

The work of Dhir *et al.* [4.5] reveals a lesser dependence on curing conditions: for concrete samples with identical strength and workability,

Table 4.5. Composition of concretes for permeability measurements (Fig. 4.15)

Mix n°	Cement	Concrete composition			
		Cement content (kg/m³)	Additions (kg/m³)	Paste content (dm³/m³)	Water content (dm³/m³)
1	PZ 35 F	271	–	277	190
2	PZ 35 F	300	–	292	195
3	PZ 35 F	300	–	277	180
4	PZ 35 F	337	–	278	169
5	HOZ 35 L	300	–	295	
6	HOZ 35 L	300	–	280	
7	PZ 35 F	240	F-ash I : 60	280,f.incl.	180
8	PZ 35 F	240	F-ash II: 60	280,f.incl.	180

Table 4.6. Method of curing of concrete in Table 4.5

Curing type	Method and time of curing				
	In mould	In air current (v=1.5m/s)	Sealed	In air conditioned room (65% RH)	concrete age
	(days)	(days)	(days)	(days)	(days)
1	1	1	–	33	35
2	1	–	–	41	42
3	1	–	2	46	49
4	1	–	28	27	56

there is little correlation between cement type and air permeability at 28 days (Fig. 4.13). As clinker content is lower in fly ash cement, Pomeroy [4.38] indicates that the initial moist curing should last much longer (Fig. 4.14). In fact, results to be found in the literature are extremely

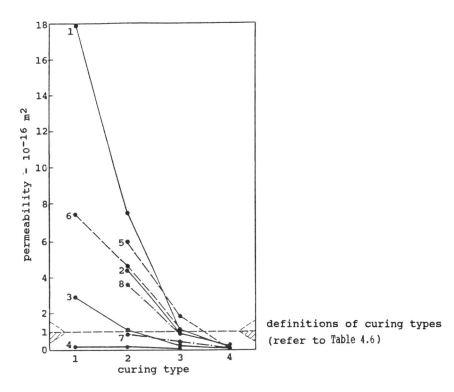

Fig. 4.12. Oxygen permeability of concrete with additions, depending on curing time and w/c ratio; cf. Tables 4.5 and 4.6 [4.36].

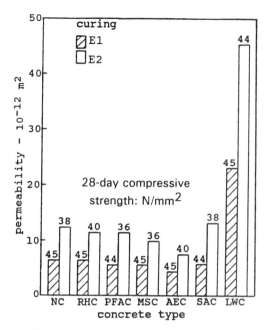

Fig. 4.13. Effect of constituent materials on air permeability of concrete with equivalent strength [4.5].

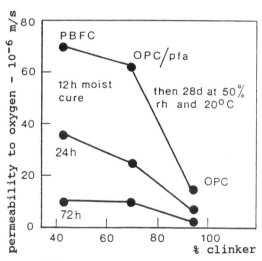

Fig. 4.14. Effect of clinker content and curing duration on the oxygen permeability of concrete [4.38].

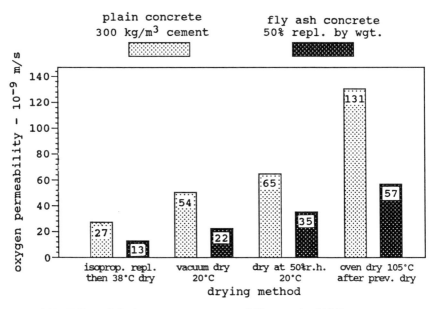

Fig. 4.15. Effect of drying on oxygen permeability results [4.35].

varied, and it is difficult to form a clear idea about the influence of mineral additions. Moreover, additives are often used to replace clinker, resulting in cement of different strengths as well as in variations in hydrate quantities at the time of testing.

The effect of additions is often analysed in terms of its activity on pore distribution. In slag-blended cement pastes, the volume of pores greater than 100 μm is negligible, even in young samples. Thus it is that, in spite of a much greater total porosity, the permeability of 70% slag cement pastes is lower than in 30% slag pastes [4.31]. On the other hand, fly ash cement pastes contain large pores when young, which diminish in time. With silica fume, pore volume is rapidly reduced. In the view of Uchikawa [4.19], the permeability of cement with additives must be determined according to such changes in pore size.

Oxygen permeability measurements carried out on concrete show the influence of fly ash additives and of specimen preconditioning [4.35]. As Fig. 4.15 reveals, if drying is carried out by using isopropanol to replace water, permeability is approximately 20–30% less than that measured by the more traditional methods of drying at 105°C. Using a variety of methods of preparation, permeability is less with fly ash. Cement pastes containing 30% natural pozzolana or fly ash are considerably less susceptible to cracking than Portland cement pastes [4.39]. Yet, when drying conditions become more severe, there does not seem to be any greater difference in oxygen permeability between the two types of concrete. We cannot however rule out the possibility that the two types of concrete may present differing initial states of micro-cracking.

4.1.4 CONCLUSIONS

The permeability of concrete, like that of other porous materials, is a function of the size distribution and the connectivity of the pores. In practice, it is not easy to analyse the influence of a particular composition parameter because its variation usually implies the modification of another one. From a practical point of view, it could seem suitable to compare permeability data for concretes of the same strength classes.

From the analysed literature it may be concluded that, for a given water/cement ratio, the basic parameter is the amount of hydrates. By considering this quantity, it is possible to analyse the influence of cement content, age and time of curing. The pore size distribution of concrete is modified by mineral additions and the permeability may be lowered if the curing is sufficient.

It should be mentioned that numerical data provided by different authors are sometimes very different because the moisture content of the tested specimens and the experimental procedures are not standardized.

4.2 Technological influences (materials and execution) on water absorption

The rate at which water is absorbed into concrete by capillary suction can provide useful information relating to the durability of concrete. The water absorption depends on a variety of parameters, which can be grouped into several categories:

- constituents characteristics,
- sample characteristics,
- stress conditions.

This reports deals only with the influence of the type and proportions of the constituents, as well as with the effect of age and curing conditions. We shall study water absorption of ordinary Portland cement (OPC) concrete in a first part without additives, and in a second part concrete with additives.

4.2.1 WATER ABSORPTION OF OPC CONCRETE WITHOUT ADDITIVES

(a) Water/cement (w/c) ratio

For concrete, the water absorption is reduced when the water/cement ratio decreases. The results of Schönlin and Hilsdorf [4.39] confirm this. They

Table 4.7. Water absorption [4.40]

Cement (kg/m³)	Water (l/m³)	w/c	Aggregate (kg/m³)	n	log(w₁)
360	162	0.45	1880	0.33	1.02
300	180	0.60	1880	0.45	1.20

Table 4.8. Water penetration results [4.41]

Cement (kg/m³)	w/c	Duration of fog curing		
		0 days	7 days	28 days
275	0.49	25	22	19
350	0.42	25	17	17
435	0.34	23	17	17

measured the weight of water absorption on cylindrical specimens sealed at the circumferential surfaces by means of a rubber ring. The cylinders were placed in a container, one surface being exposed to water. They present their results by the following expression [4.40]:

$$w = w_1 t^n$$

where w_1 = amount of water absorbed during the first hour
 n = exponent
 t = time

The values of n and $\log(w_1)$ are shown in Table 4.7. Haque [4.41] also verifies these results. He immersed concrete cylinders after different curing durations in a bucket of water, and after 2 h of immersion the depth of water penetration was measured (Table 4.8).

The reduction of water absorption for decreasing water/cement ratio is more significant in the work of Dhir *et al.* [4.42]. They measured the initial surface absorption (ISA) at 10 min for concretes of different water/cement ratios and for various initial water-curing durations (Fig. 4.16).

The data obtained by Hall [4.43] concern a large range of w/c ratios (Table 4.9). He expresses the sorptivity in mm/min$^{1/2}$: except for w/c = 0.4, the water absorption increases with the water/cement ratio.

(b) Curing

Effect of wet curing The effect of a moist curing for concretes manufactured with different water/cement ratios has been studied by Haque [4.41]. The water penetration is more important when no fog curing was provided

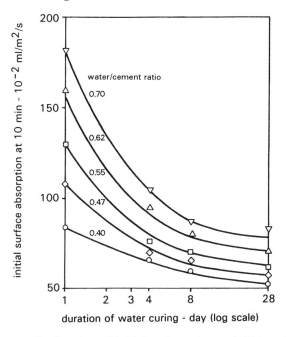

Fig. 4.16. Influence of *w/c* ratio and initial moist curing on initial surface absorption (test age: 28 days) [4.42].

(Table 4.8). Kelham [4.44] confirms this result. He measured the sorptivity S in mm/h$^{1/2}$ of a 28-day old concrete (w/c = 0.59) for two curing conditions:

- air stored: $S = 20.0$ mm/h$^{1/2}$,
- water stored: $S = 14.3$ mm/h$^{1/2}$.

Influence of curing duration   The duration of curing also plays an important role, as shown in Table 4.10 [4.39]. Water absorption increases as the duration of water curing decreases.

Table 4.9. Variation of sorptivity with water/cement ratio [4.43]

Composition	w/c	Porosity	Sorptivity (mm/min$^{1/2}$)
	0.40	–	0.480
Cement : sand : aggregate	0.50	0.113	0.250
	0.60	–	0.290
1 : 2 : 4	0.70	0.139	0.290
	0.80	0.141	0.350
	0.90	0.132	0.360

Table 4.10. Effect of duration of curing [4.39]

w/c	Duration of curing (days)	log(w_1)	n
0.45	1	1.02	0.33
	3	0.82	0.27
	7	0.66	0.27
0.60	1	1.20	0.45
	3	0.87	0.33
	7	0.68	0.32

Parrott's work [4.45] verifies these results (Fig. 4.17). The reduction of water absorption with increasing curing time is also found by Lydon *et al.* [4.46]. The results are shown in Table 4.11.

Table 4.11. Effect of curing on water absorption [4.46]

Cement (kg/m³)	Water (kg/m³)	Curing history	ISA (ml/m²s) × 10⁻²		
			10 min	30 min	1 h
330	170	7 days in air	12	5	2
330	170	1 day in water + 6 days in air	10	4.5	1

However, it seems that the water absorption decreases regularly with an initial curing time of up to 7 days (Table 4.10), but between 7 and 28 days curing, the improvement is pronounced (Table 4.8). This result is confirmed by Dhir *et al.* [4.42] as shown in Fig. 4.16: the ISA for concretes prepared

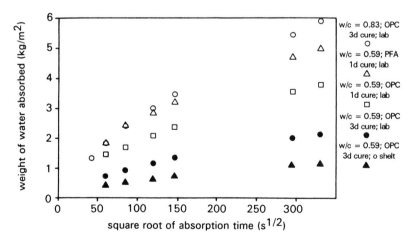

Fig. 4.17. Influence of curing on water absorption [4.45].

Table 4.12. Influence of duration of curing on water absorption [4.46]

Duration of moist curing	Water absorption (kg/m^2) after:		
	1 h	4 h	24 h
1 day	1.46	2.07	3.58
3 days	0.77	1.19	1.99
28 days	0.71	1.08	1.88

with different w/c ratios reaches a limit when the curing time increases.

Parrott [4.46] verifies that for OPC concrete, long-term curing had only a small effect (Table 4.12).

Effect of exposure conditions Parrott also studies the effect of exposure conditions upon absorption. Water absorption is measured on OPC concrete exposed to different conditions (Table 4.13). The results are shown in Fig. 4.18. It was evident that absorption was much smaller in wetter concretes. The effect was particularly significant when rain could fall on the exposed concrete surface.

Gowripalan *et al.* [4.47] have carried out water absorption measurements on concrete that has been cured in different ways:

● cured under burlap;
● cured under plastic sheet;
● a poor-quality clear translucent type of curing compound; and
● non-cured specimens.

Figure 4.19 presents the coefficient of absorptivity, which is the water absorbed per unit area within a given period of exposure. Poor curing

Table 4.13. Exposure conditions for data in Fig. 4.18 [4.46]

	Relative humidity (%)	Temperature (°C)
Laboratory	58 ± 3	20 ± 1
Office	45	15 to 25
Outside sheltered from rain	58 to 90	6 to 21
Outside vertical exposed surface	58 to 90	6 to 21
Outside horizontal exposed surface	58 to 90	6 to 21

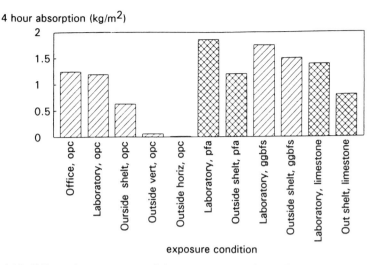

Fig. 4.18. Effect of exposure condition upon water absorption after 1.5 years of exposure (3 days moist curing 0.59 *w/c* ratio) [4.46].

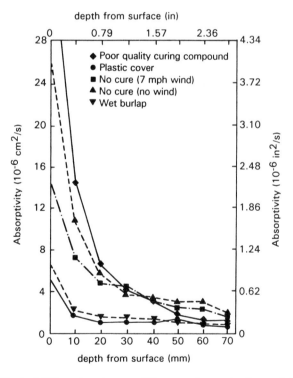

Fig. 4.19. Effect of different curing conditions on change in absorptivity with depth for samples made of regular sand cured at 22% RH for 5 days [4.47].

Table 4.14. Curing conditions for data in Table 4.15
[4.48]

Environment	Conditions
E1	Water 5°C
E2	Water, 23°C
E3	3 days of water, 23°C; thereafter air, 23°C, 55% RH
E4	Waster, 50°C
E5	24 hours of air, 5°C, 95% RH; 27 days of air, 10°C, 55% RH; thereafter air, 23°C, 55% RH
E6	Air, 10°C, 55% RH
E7	Air, 23°C, 55% RH
E8	Air, 50°C, 55% RH

resulted in an extremely high absorptivity near the surface, and the effect of curing is more pronounced within a surface near region of approximately 30 mm thickness. This region constitutes the cover zone for most reinforced concrete structures. Poor-quality concrete in this region will enhance corrosion of the steel reinforcement.

Effect of curing temperature The work of Dhir *et al.* [4.48] reveals a dependence on curing temperature. One grade of concrete (35 N/mm^2) was cured at temperatures varying from 5 to 50°C in water or air (Table 4.14) and was tested at the age of 28 days for its absorption properties. The results obtained are given in Table 4.15. This table shows that both absorption parameters (ISA-10 and *n*) generally improve with curing temperature.

(c) Aggregates

Few studies have been made on the influence of aggregates on water absorption. Marchese *et al.* [4.49] have measured the rate of water absorption for mature pastes and mortars made with two types of cement (OPC and pozzolanic cement with the same clinker) and two types of sand (normal sand and monogranular sand). Compositions are given in Table 4.16. Results in Table 4.17 show that the OPC paste (*w/c* = 0.5) was able to absorb the greatest amount of water and at the highest rate. The corresponding mortar, made with well-graded sand (water/cement/sand = 0.5/1/3) showed a much lower sorptivity. At the same cement/sand ratio the reduction was shown to depend on the content of paste filling the voids between grains, and then on the sand grading. Using 'monogranular' fraction with 0.5–1 mm grains of the same sand the greater amount of paste available for filling the higher volume between the grains offers easier inflow of water in the resulting mortar.

Table 4.15. Effect of curing conditions on water absorption [4.48]

Concrete mix	Curing	Deviations			
		From corresponding E1/E5*		From OPC/E²	
		ISA–10 (ml/m²s × 10⁻²)	n × 10²	ISA–10 (ml/m²s × 10⁻²)	n × 10²
Water curing					
OPC: M2	E1	0	0	+58	−11
	E2	−58	−11	0	0
	E4	−80	−1	−22	−1
OPC PFA15: M7	E1	0	0	−12	+6
	E2	+4	+3	−8	+9
	E4	−9	+12	−22	+18
OPC PFA15: M12	E1	0	0	−3	+5
	E2	−6	+13	−9	+18
	E4	−31	+16	−34	+21
Air curing					
OPC: M2	E5	0	0	+157	−4
	E6	−29	−1	+157	−4
	E7	−2	−5	+155	+1
	E8	+2	+2	+159	−2
OPC PFA15: M7	E5	0	0	+131	−5
	E6	−50	+2	+81	−3
	E7	−40	0	+91	−5
	E8	−79	+20	+52	+17
OPC PFA30: M12	E5	0	0	+197	−10
	E6	−92	+5	+105	−5
	E7	−75	+10	+122	0
	E8	−100	+25	+97	−15

*Water curing with respect to E1, air curing with respect to E5.

In his review on water sorptivity of mortars and concretes Hall [4.43] finds that sand quantity variation has no influence on water absorption (Table 4.18).

(d) Workability

The only results concerning concrete workability have been obtained by Dhir *et al.* [4.42]. They have made series of tests on concrete mixes with a constant *w/c* ratio of 0.55, but with different workabilities obtained by varying either the water content or by using a plasticizing admixture of the dispersing type. Figure 4.20 shows that ISA increases with increasing workability, at a significantly greater rate when the increase in workability is associated with an increase in the water content (and hence cement content to maintain a constant *w/c* ratio) than when the increase in work-

Table 4.16. Composition of mixes in Table 4.17 [4.49]

	OPC	Pozzolanic cement	Normal sand	Monogranular sand
1 paste	1	–	–	–
2 paste	–	1	–	–
3 mortar	1	–	3	–
4 mortar	–	1	3	–
5 mortar	1	–	–	3
6 mortar	–	1	–	3

Table 4.17. Effect of aggregates on sorptivity [4.49]

	Bulk density (kg/m^3)	Porosity $(\%)$	Sorptivity $(mm/min^{1/2})$
1 paste	1592	35.04	1.302
2 paste	1449	41.96	0.385
3 mortar	2229	13.91	0.090
4 mortar	2194	15.35	0.067
5 mortar	2134	16.34	0.139
6 mortar	2103	17.70	0.116

Table 4.18. Effect of sand on sorptivity [4.43]

Composition	w/c	Porosity	Sorptivity $(mm/min^{1/2})$
cement:sand:aggregate	0.60	–	0.290
1:2:4	0.80	0.141	0.350
cement:sand:aggregate	0.60	–	0.290
1:3:4	0.80	–	0.310

ability is obtained by the use of plasticizing admixtures. For Dhir *et al.* the fluid nature of high-workability superplasticized concrete, together with the possible influence of superplasticizer on the physical properties of the gel [4.50], explains its slightly higher ISA values. However, it is clear from Fig. 4.20 that this type of concrete has a lower ISA value than concrete of the same workability without superplasticizer. It is shown that for the same *w/c* and workability, superplasticizer can be employed effectively to reduce the absorptivity of concrete.

(e) Compaction

Hall [4.43] presents some results on the influence of the compaction of the sample (Table 4.19). Sorptivity decreases for the same concrete when

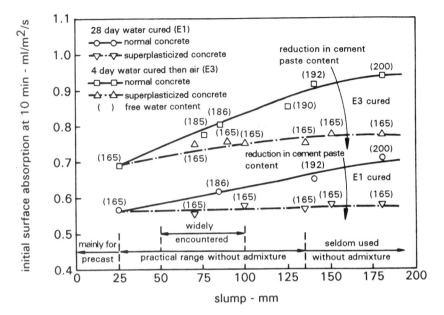

Fig. 4.20. Effects of concrete slump on its initial surface absorption [4.42].

tamping is prolonged.

(f) Age

As capillary porosity decreases with the progression of hydration reaction, so concrete water absorption decreases with age. Marchese *et al.* [4.49] have worked on the water absorption of pastes and mortars with an age of 8 months and 1 year, respectively. The absorption of the paste

Table 4.19. Effect of compaction on sorptivity [4.43]

Composition	w/c	Porosity	Sorptivity (mm/min$^{1/2}$)
	0.40	–	0.480
cement:sand:aggregate	0.50	0.113	0.250
	0.60	–	0.290
1:2:4	0.70	0.139	0.290
	0.80	0.141	0.350
normal tamping	0.90	0.132	0.360
	0.40	–	0.094
cement:sand;aggregate	0.50	–	0.120
	0.60	0.113	0.170
1:2:4	0.70	0.129	0.130
	0.80	0.133	0.180
prolonged tamping	0.90	0.138	0.130

Table 4.20. Effect of age on sorptivity [4.49]

Composition	8 months		1 year	
	Porosity (%)	Sorptivity (mm/min$^{1/2}$)	Porosity (%)	Sorptivity (mm/min$^{1/2}$)
Paste	35.04	1.302	33.95	0.485
Mortar with normal sand	13.71	0.090	11.01	0.029
Mortar with monogranular sand	16.34	0.139	14.54	0.088

and mortars was lower for the 1-year-old samples (Table 4.20).

4.2.2 WATER ABSORPTION OF CONCRETE WITH ADDITIVES

Many studies have been made on the influence of additives on concrete water absorption. However, water absorption measurements carried out on concrete do not clearly show the influence of additives, because some authors compare concretes of equal 28-day strength; other work deals with equal-workability concretes or equal *w/c* ratio. So the influence of additives was studied in three parts corresponding to these three types of comparison.

(a) Concretes of equal 28-day strength

In work published in 1987, Dhir *et al.* [4.42] measured water absorption by ISA tests on concrete mixes using different constituent materials (Table 4.21) but having similar workability (slump 75 mm) and a 28-day design strength of 45 N/mm^2 after normal curing. The results are shown in Fig. 4.21.

The surface absorption of the OPC and pulverized fuel ash (PFA) mixes

Table 4.21. Mixes used for tests in Fig. 4.21 [4.42]

Concrete	Description
OPC	OPC control N55
RHPC	Rapid-hardening cement
PFA	OPC with PFA; C + F = 390 kg/m^3 and F/(C + F) = 0.27
MS	OPC with 16% microsilica
AEA	OPC with 5% air-entrainment
SA	Superplasticizing admixture

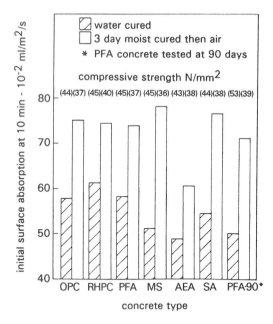

Fig. 4.21. Comparison of ISA for various types of concrete tested at 28 days [4.42].

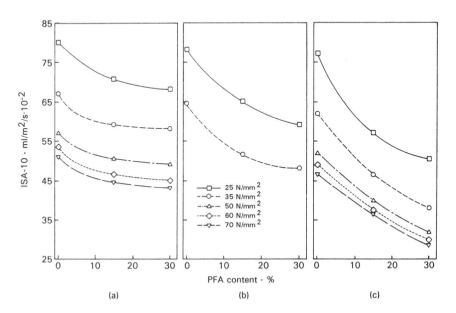

Fig. 4.22. Effect of PFA on ISA-10; curing E2: (a) 28 days, (b) 90 days, (c) 180 days [4.42].

is similar despite the modification of the pore structure caused by PFA, showing that PFA can effectively replace a part of the cement in concrete without adversely affecting the concrete absorption properties. The water-cured microsilica concrete is less absorptive than the OPC mix, but for the other curing methods the two concretes have similar absorption, owing to the slower rate of initial hydration of the microsilica concrete. This indicates that microsilica concrete suffers more from poor curing than OPC concrete. However, owing to their high pozzolanic activity, there is a significant reduction in the absorption of microsilica and PFA concretes when they are continuously moist-cured for a long period. Despite the greater fineness of rapid-hardening Portland cement (RHPC), concrete of the same strength made with it gave similar ISAT values to those of a OPC mix. For Dhir *et al.* the lower absorption of air-entrained concrete is due to the combined effect of reduction in water/cement ratio and water content, improved consistency of the mix and the fact that the air bubbles are not interconnected.

In another paper published in 1991, Dhir *et al.* [4.48] deal more precisely with the influence of PFA in combination with OPC and RHPC. Two series of mixes were designed with OPC and RHPC, covering a range of 28-day standard strengths from 25 to 70 MPa. All mixes were proportioned for 75 ± 25 mm slump. The mix proportions are given in Table 4.22. The curing conditions used are listed in Table 4.14.

The ISA-10 results obtained for concrete mixes of equal 28-day strength with E2 (water 23°C) curing are plotted in Fig. 4.22. All PFA concrete mixes absorbed water at a lower rate than the corresponding OPC concrete mixes. This effect increases with the PFA content at a decreasing rate up to 30% and with a duration of moist-curing up to 180 days. Although for a given 28-day design strength PFA concretes have a lower $w/(c + F)$ ratio than OPC concretes (Table 4.22), for Dhir *et al.* this alone does not account for their better performance, as can be seen from Fig. 4.23. The reduction in surface absorption of 28-day PFA concrete is thought to be due to modifications in the microstructure such as pore discontinuity caused by PFA and any pozzolanic reaction.

The results obtained with curing in air at 23°C and 55% RH (E7) are shown in Fig. 4.24. It can be seen that for both OPC and OPC/PFA concrete ISA-10 is substantially higher with air curing than with water curing. It is also shown that the beneficial effect of PFA with E2 curing remains with E7 curing. The effect of low-temperature air curing (E5) on the water absorption is compared with E7 curing in Table 4.23. This shows a general reduction in the quality of concrete with E5 curing. The effect of PFA varies with the replacement level. At 15% it reduces the adverse effect of E5 curing on OPC concrete whereas at 30% replacement it lowers the quality below that observed for OPC concrete.

One strength grade concrete (35 MPa) was cured at temperatures

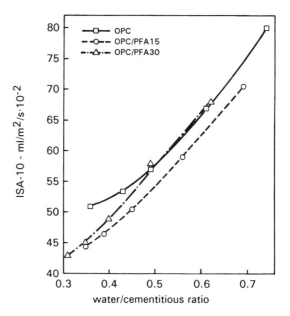

Fig. 4.23. Effect of *w/c* ratio on ISA-10; curing E2, 28 days [4.22].

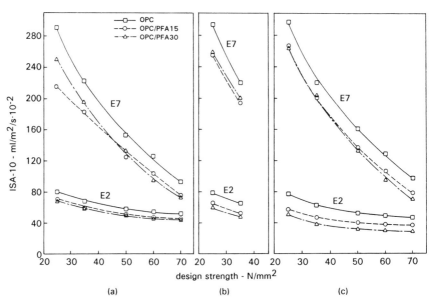

Fig. 4.24. Effect of air-curing (E7) on ISA-10: (a) 28 days, (b) 90 days, (c) 180 days [4.48].

Table 4.22. Mix proportions for tests in Fig. 4.22 [4.48]

Concrete mix			Mix proportions (kg/m³)					
Code	Strength (N/mm²) Design	Measured	OPC/RHPC (c)	PFA (f)	c+f	Water (w)	Aggregate	w/c+f
Series 1 OPC:								
M1	25	24.0	250	–	250	185	1960	0.74
M2	35	35.0	305	–	305	185	1915	0.61
M3	50	52.0	380	–	380	185	1850	0.49
M4	60	62.0	435	–	435	185	1805	0.43
M5	70	71.0	510	–	510	185	1745	0.36
OPC PFA15:								
M6	25	25.5	215	40	255	175	1970	0.69
M7	35	36.5	265	45	310	175	1925	0.56
M8	50	52.0	330	60	390	286	1850	0.45
M9	60	62.0	380	65	445	175	1805	0.39
M10*	70	71.5	430	75	505	175	1750	0.35
OPC PFA30:								
M11	25	25.0	190	80	270	170	1955	0.63
M12	35	36.0	240	105	345	170	1885	0.49
M13	50	52.0	295	125	420	170	1820	0.40
M14*	60	61.0	345	145	490	170	1755	0.35
M15*	70	71.5	375	165	540	170	1705	0.31
Series 2 RHPC:								
M18	25	27.0	235	–	235	190	1965	0.81
M19	35	34.5	270	–		190	1935	0.70
M20	50	50.5	345	–	345	190	1870	0.55
M21	60	59.0	395	–	395	190	1830	0.48
M22	70	68.5	440	–	440	190	1795	0.43
RHPC PFA15:								
M23	25	26.5	205	35	240	180	1975	0.75
M24	35	35.5	240	45	285	180	1935	0.63
M25	50	53.0	300	55	355	180	1870	0.51
M26	60	61.0	350	60	410	180	1825	0.44
M27	70	71.5	395	70	465	180	1775	0.38
RHPC PFA25:								
M28	25	24.5	185	60	245	175	1975	0.71
M29	35	34.5	220	75	295	175	1930	0.59
M30	50	50.0	290	95	385	175	1845	0.45
M31	60	60.5	335	110	445	175	1800	0.39
M32*	70	690.	365	120	485	175	1760	0.36

*Superplasticizer was used to achieve the required slump.

Table 4.23. Effect of curing conditions on water absorption of PFA concrete [4.48]

Concrete standard strength (N/mm^2)	Deviation from corresponding measurements with E7 curing					
	28 days		90 days		180 days	
	ISA–10 ($ml/m^2s \times 10^{-2}$)	$n \times 10^2$	ISA–10 ($ml/m^2s \times 10^{-2}$)	$n \times 10^2$	ISA–10 ($ml/m^2s \times 10^{-2}$)	$n \times 10^2$
OPC:						
M1 25	+102	–8	+142	–5	+164	–8
M2 35	+14	–6	+39	–4	+59	–3
M3 50	–17	–4	–	–	+10	+3
OPC PFA15:						
M6 25	+99	+2	+65	–2	+58	–3
M7 35	+14	–1	+16	0	+20	+2
M8 50	–8	–1	–	–	+17	–3
OPC PFA30:						
M11 25	+155	–5	+188	–6	+207	–7
M12 35	+60	–3	+76	–3	+90	–4
M13 50	+32	–4	–	–	+64	–3

varying from 5 to 50°C in water or air, and was tested at the age of 28 days (Table 4.24). This shows that absorption improves with curing temperature, and the effect is well defined in the air-cured PFA concrete. The ISA-10 results in Table 4.25 show that the surface quality of concrete decreases with decreasing duration of wet curing. OPC concrete is more affected than PFA concrete.

Parallel studies were carried out on concrete mixes made with RHPC. For water curing (E2) the results obtained are presented in Fig. 4.25 as the deviations of the ISA-10 values from those obtained for the 28-day standard curing. It is shown that at the test age of 28 days RHPC concrete has higher ISA values than OPC concrete. This is probably a direct result of the higher mixing water content and w/c ratio in RHPC concrete than in the corresponding OPC concrete. However, this weakness of RHPC is well compensated by the use of PFA. For air curing (E7), the use of RHPC decreases ISA-10, and the PFA concrete mixes show a considerable further improvement (Fig. 4.26(a)). Figure 4.27 shows the influence of E5 curing. For all strengths the RHPC mixes had lower ISA-10 values. Although smaller differences are shown at the age of 180 days, it should be noted that a comparison with the OPC/E5-180 mixes shows that the RHPC mixes maintain their advantage (Fig. 4.26(b)). A replacement level of 15% PFA was found to optimize the performance of both OPC and RHPC in this environment, but in all cases the corresponding RHPC/PFA mixes were the best.

Haque [4.41] reveals the influence of curing on water absorption of PFA

Table 4.24. Effect of curing conditions on ISA of PFA concretes [4.48]

Concrete mix	Curing	Deviations			
		From corresponding E1/E5*		From OPC/E2	
		ISA–10 (ml/m²s × 10⁻²)	n × 10²	ISA–10 (ml/m²s × 10⁻²)	n × 10²
Water curing					
OPC: M2	E1	0	0	+58	−11
	E2	−58	−11	0	0
	E4	−80	−1	−22	−1
OPC PFA15:	E1	0	0	−12	+6
M7	E2	+4	+3	−8	+9
	E4	−9	+12	−22	+18
OPC PFA30:	E1	0	0	−3	+5
M12	E2	−6	+13	−9	+18
	E4	−31	+16	−34	+21
Air curing					
OPC:M2	E5	0	0	+157	−4
	E6	−29	−1	+128	−3
	E7	−2	−5	+155	+1
	E8	+2	+2	+159	−2
OPC PFA15:	E5	0	0	+131	−5
M7	E6	−50	+2	+81	−3
	E7	−40	0	+91	−5
	E8	−79	+20	+52	+17
OPC PFA30:	E5	0	0	+197	−10
M12	E6	−92	+5	+105	−5
	E7	−75	+10	+122	0
	E8	−100	+25	+97	+15

*Water curing with respect to E1, air curing with respect to E5.

Table 4.25. Effect of curing conditions on ISA of PFA concretes [4.48]

Concrete mix	Curing	Deviations			
		From corresponding E2		From OPC/E2	
		ISA–10 (ml/m²s × 10⁻²)	n × 10²	ISA–10 (ml/m²s × 10⁻²)	n × 10²
OPC control:	E2	0	0	0	0
M2	E3	+27	+10	+27	+10
	E7	+155	+1	+155	+1
OPC PFA15:	E2	0	0	−8	+9
M7	E3	+30	−2	+22	+7
	E7	+99	−14	+91	−5
OPC PFA30:	E2	0	0	−9	+18
M12	E3	+41	−10	+32	+8
	E7	+131	−18	+122	0

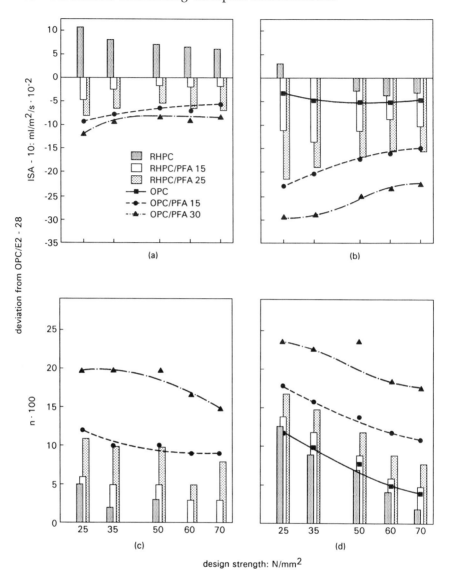

Fig. 4.25. Comparison of RHPC and OPC; curing E2: (a) 28 days (ISA-10), (b) 180 days (ISA-10), (c) 28 days (n), (d) 180 days (n) [4.48].

concretes. The various concrete mixes used in his investigation are given in Table 4.26. Three grades of plain concrete with nominal 28-day strengths of 30, 40 and 50 MPa were prepared.

The results of water penetration (mm) during 2 h are shown in Table 4.27. The depth of water penetration in PFA concrete is lower than in the

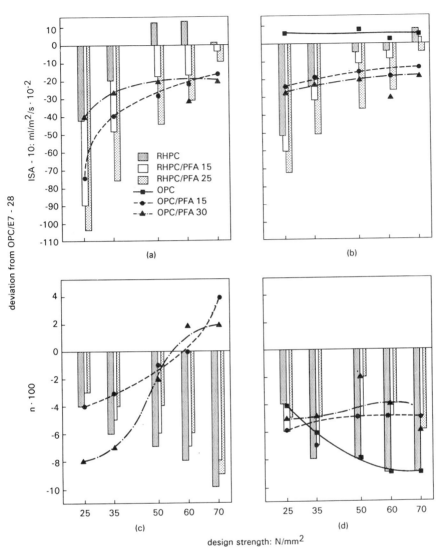

Fig. 4.26. Comparison of RHPC and OPC; curing E7: (a) 28 days (ISA-10), (b) 180 days (ISA-10), (c) 28 days (n), (d) 180 days (n) [4.48].

corresponding OPC concrete when mixes were previously fog-cured for 7 and 28 days.

In his thesis, Balayssac [4.51] compares water absorption for different cement concretes with the same 28-day strength (35 MPa). Five cements are used:

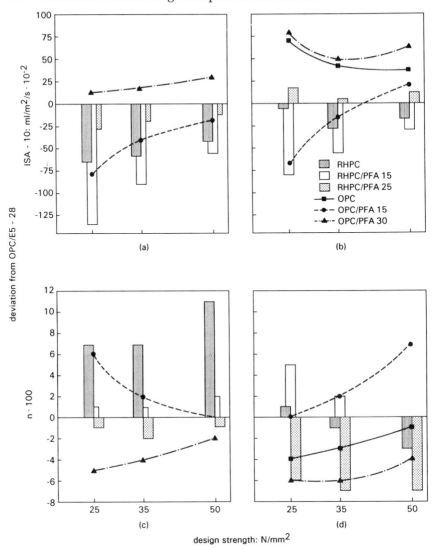

Fig. 4.27. Comparison of RHPC and OPC; curing E5: (a) 28 days (ISA-10), (b) 180 days (ISA-10), (c) 28 days (n), (d) 180 days (n) [4.48].

- an ordinary Portland cement, CPA HP;
- an ordinary Portland cement (79%) with limestone filler (19%), CPJf 45;
- an ordinary Portland cement with a 55 MPa nominal strength, CPA 55;
- a rapid-hardening ordinary Portland cement, CPA HPR;
- an ordinary Portland cement with ground granulated blastfurnace slag (GGBS>80%), CLK 45.

Table 4.26. Mixes used for tests in Table 4.27 [4.41]

Mix	Cement(c)+ fly ash(f) (kg/m³)	w/(f+c)	Slump (mm)
30–0	275	0.49	30
30–30	(192.5+82.5)	0.45	30
40–0	350	0.42	30
40–30	(245+105)	0.39	40
50–0	435	0.34	50
50–30	(304.5+130.5)	0.37	45

Table 4.27. Water penetration (mm) after 2 h for PFA concretes [4.41]

	Duration of prior fog curing		
Mix	Zero	7 days	28 days
30–0	25	22	19
30–30	37	20	17
40–0	25	17	17
40–30	34	21	15
50–0	23	17	17
50–30	28	15	13

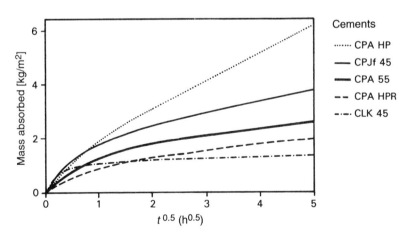

Fig. 4.28. Water absorption of 35 MPa 28-day strength concrete cured 3 days (age 90 days) [4.51].

Table 4.28. Mix proportions for mixes in Fig. 4.29 [4.52]

Concrete mix	Mix proportions (kg/m³)			ThamesValley			Free water/ cement ratio: maximum	Average slump (mm)	Cube results (N/mm²)				
	Cement	PFA	GGBS	20–10	10–5	< 5			3 days	4 days	7 days	14 days	28 days
OPC	315	–	–	680	340	825	0.55	67.5	15.3	19.0	36.6	37.4	47.4
PFA	240	100	–	680	340	825	0.55	77.5	11.5	–	28.5	34.5	42.2
GGBS	160	–	155	680	340	825	0.55	80.0	13.3	–	22.6	31.3	33.4

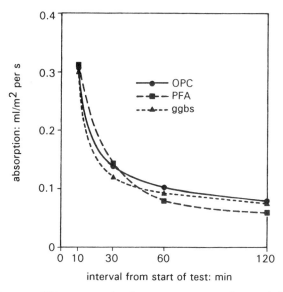

Fig. 4.29. Average ISA test results for concrete cores from reinforced concrete tank walls [4.52].

The results shown in Fig. 4.28 confirm the influence of the cement on concrete water absorption.

(b) Concretes of equal w/c ratio

Thomas *et al.* [4.52] have compared the properties of cores taken from the walls of a 2.5-year-old reinforced concrete tank manufactured with OPC, PFA and GGBS of equal strength grade. The mix designs of the three concretes are given in Table 4.28. ISA measurements are shown in Fig. 4.29. The results indicate only small differences between the average surface absorption of concretes.

Lydon *et al.* [4.46] have compared OPC and PFA concretes for different curing conditions (Table 4.29). For the mixes with a total content of cementitious materials of 300 kg/m^3 1-day water curing after demoulding has resulted, on average, in an improvement of 124% and 43% in the ISA values of PFA concrete and OPC concrete respectively. For the other mixes, extending the air curing from 7 days to 100 days has resulted, on average, in increase of 78% and 152% in the ISA values of PFA concrete and OPC concrete respectively. This suggests that PFA concrete responds more favourably to initial moist curing and suffers less during prolonged dry air exposure.

Parrott [4.45] has studied the influence of additives on water absorption concretes. Figure 4.30 reveals for different water/cement ratios that the OPC concretes generally exhibited the lowest water absorption at a

Table 4.29. Effect of curing conditions on ISA for PFA concretes [4.46]

OPC content (kg/m³)	PFA content (kg/m³)	Nominal water content (kg/m³)	Curing history	ISA (10⁻²ml/m²s)		
				10 min	30 min	1
330	–	170	7 days	12	5	2
231	99	170	in air	22	11	5
330	–	170	1 days in water	10	4.5	1
231	99	170	then 6 days in air	13	6.5	1.5
450	–	165	7 days	8.5	3.5	0.5
315	135	165	in air	13	6	2.5
450	–	165	100 days	14.5	6.5	2
315	135	165	in air	20	10	5.2

*Air drying was at 20°C and 68% RH immediately after demoulding at 1 day, apart from that water-cured for 1 day.

given water/cement ratio. The replacement of 5% of the OPC with limestone filler caused a small increase in absorption. The beneficial effects of curing upon the resistance of concrete to water absorption are demonstrated in Table 4.30. Curing beyond 3 days reduces the absorption levels for concretes made with PFA or GGBS but for OPC long-term curing has only a small effect. When OPC was partially replaced with PFA or GGBS it was necessary to increase the curing period from 3 to 28 days in order to achieve similar absorption results.

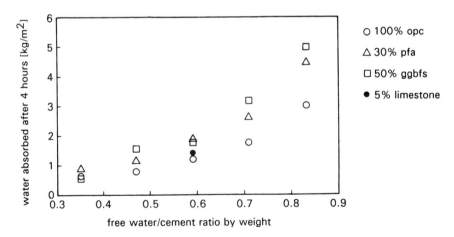

Fig. 4.30. Effect of *w/c* ratio and cement type upon water absorption after 1.5 years of laboratory exposure (3 days moist curing) [4.45].

Table 4.30. Effect of curing on water absorption [4.45]

Moist cure (days)	100% OPC			30% PFA			50% ggbfs		
	1 h	*4 h*	*24 h*	*1 h*	*4 h*	*24 h*	*1 h*	*4 h*	*24 h*
1	1.46	2.07	3.58	1.84	2.85	4.71	1.39	2.40	4.70
3	0.77	1.19	1.99	1.31	1.86	2.70	1.08	1.75	2.97
28	0.71	1.08	1.88	0.95	1.34	2.09	0.97	1.37	1.92

Marchese *et al.* [4.49] have worked on paste and mortar mixes made with OPC and PFA. Compositions are given in Table 4.16. Figures 4.31 and 4.32 show water absorption results for 8-month-old and 1-year-old samples.

When using pozzolanic cement with the same water/cement ratio, a lower water absorption capacity was observed for both the paste and the two mortars, relative to the corresponding OPC mixes. After 1 year the absorption capacity was reduced for all the samples. A more noticeable reduction for OPC mortars relative to the corresponding pozzolanic mortars was observed.

(c) Concretes of equal workability

Kelham [4.44] studied the influence of curing on three mixes: an OPC concrete, an OPC with 25% PFA concrete and an OPC with 60% GGBS concrete. The concretes were based on a 1 : 2.5 : 3.5 (cement : sand : coarse aggregate) mix with the water content adjusted to give a slump of 60 mm. The results are shown in Fig. 4.33 for water-cured concretes and

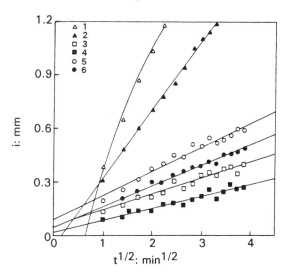

Fig. 4.31. Cumulative water absorption for 8-month old samples from Table 4.16 [4.49].

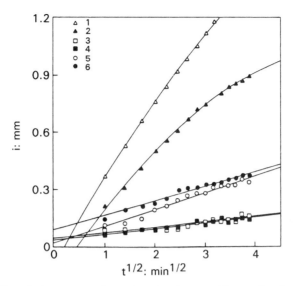

Fig. 4.32. Cumulative water absorption for 1-year old samples from Table 4.16 [4.49].

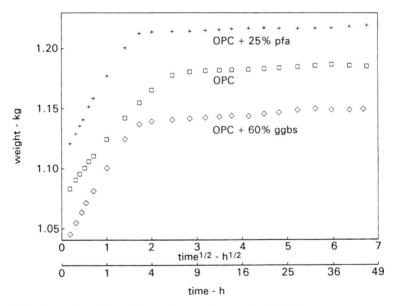

Fig. 4.33. Experimental results: water-cured concretes [4.44].

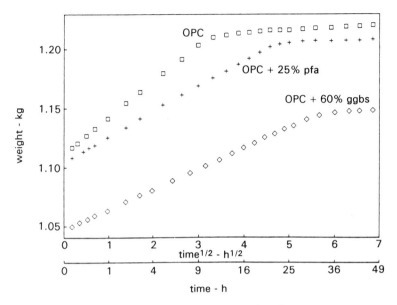

Fig. 4.34. Experimental results: air-cured concretes [4.44].

in Fig. 4.34 for air-cured concretes. Figure 4.33 demonstrates the beneficial effect of PFA, but Fig. 4.34 shows the sensitivity of the concretes containing PFA or GGBS to poor curing.

4.2.3 CONCLUSIONS

In conclusion, for every type of concrete, water absorption is reduced when:

● the water/cement ratio decreases;
● the concretes are wet-cured at young ages;
● the duration of moist curing increases;
● the curing temperature decreases;
● the aggregates have a large size range;
● compaction quality is better;
● a superplasticizer is employed;
● the age of the samples increases.

The influence of additives is not clearly pointed out in the literature. However, at equal water/cement ratio, the use of PFA and other additives does not seem to improve concrete water absorption. But for equal 28-day strength concretes, the replacement of a part of the cement with PFA reduces water absorption.

4.3 Environmental parameters influencing permeability

4.3.1 INTRODUCTION

This review is largely concerned with the effects of moisture content and the internal relative humidity upon gas permeation and water absorption in concrete, particularly in the zone adjacent to an exposed surface. Transport of gases, liquids and ions through this surface zone (**cover concrete**) is often a critical factor for concrete durability. The effects of temperature and carbonation upon permeation are also considered, although lack of data precludes detailed discussion.

The term **permeation** is used to include mass transport due to pressure gradients, concentration gradients and capillary suction. Permeation in concrete is strongly dependent upon the volume and connectivity of the larger (capillary) pores in the hardened cement paste matrix.

4.3.2 EFFECT OF MOISTURE UPON GAS TRANSPORT

Numerous publications indicate that the air permeability of concrete increases significantly as moisture is removed [4.5, 4.35, 4.53–4.68]. This is due to the increased volume and connectivity of channels available for permeation. Data from selected publications will be reviewed and then a summary of results will be presented.

Ujike *et al.* [4.53, 4.54] uniaxially dried small concrete samples ($150 \times 150 \times 60$ mm) at 35, 60 and 85% relative humidity and 20°C, and monitored the changes of air permeability during drying. The samples reached approximate moisture equilibrium after about 3 months of drying. Results for concretes of different water/cement ratios (Fig. 4.35) show that the increased permeability is closely related to the loss of moisture (the term **porosity** is used by Ujike *et al.* to denote the volume of moisture lost as a proportion of the concrete volume). The relationship between permeability and 'empty porosity' does not seem to be greatly affected by the relative humidity of drying, closeness to moisture equilibrium or water/cement ratio.

Jonis and Molin [4.55] reported a decrease in air permeation time (an increase of permeability) with drying for concretes of different strengths and having 8 mm or 32 mm maximum size aggregates. The permeation time (Figg test) results are plotted against the relative humidity in the concrete in Fig. 4.36.

Parrott and Chen have obtained results for air permeation through cover concrete using a pressurized cavity method [4.56–4.59]. Relative humidity was measured in the cavity prior to each permeability test. Figure 4.37 shows that relative humidity results mirror the permeability results when the concrete surface is dried and wetted. The higher values of perme-

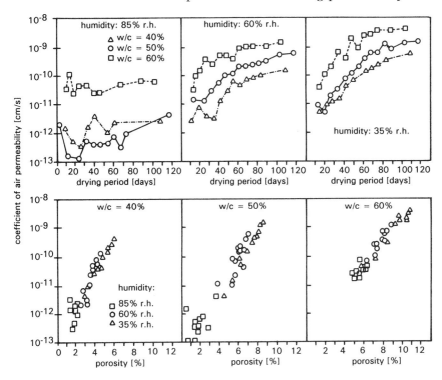

Fig. 4.35. Air permeability and moisture loss for concretes with 0.4, 0.5 and 0.6 water/cement ratios, dried at 35, 60 and 85% relative humidities; Ujike *et al.* [4.53, 4.54].

ability for cavities 1 and 2 were associated with cracks in the concrete caused by deliberate alkali–silica reaction. Air permeation through cracks was observed as large bubbles when liquid soap solution was applied to the concrete surface adjacent to the pressurized cavity.

The cracks did not seem to greatly affect the rate of water vapour diffusion. Results for uncracked concretes of different water/cement ratios are shown in Figs 4.38 and 4.39. There was a broad relationship between permeability and weight (i.e. water) loss of concrete cubes that were dried and wetted (Fig. 4.38). The hysteresis may be partly due to the test specimen geometry and to moisture gradients. The same permeability results plotted against cavity relative humidity (Fig. 4.39) exhibit hysteresis and greater differences between concretes. The results in Fig. 4.40 were obtained using the pressurized cavity method on a sheltered, externally exposed column. Long-term variations of cavity relative humidity between 40 and 70% had little effect upon the air permeability of the cover concrete. The consistently high value of permeability for cavity 3 was due to local cracking.

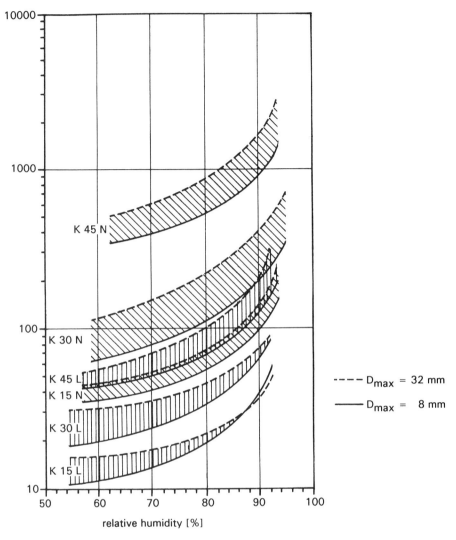

Fig. 4.36. Air permeation time plotted against relative humidity in concretes of different strengths; Figg test [4.55].

Air permeability measurements in the laboratory are often preceded by some form of oven drying to ensure a standardized test condition and significant permeation rates. Oven drying was generally found to increase gas permeability significantly [4.5, 4.35, 4.60–4.64]. The results of Nagataki [4.65] indicate that air permeability is increased with oven drying at 50°C due to loss of evaporable water and emptying of pores. Oven drying at 100°C or 150°C caused an additional increase of air permeability due to

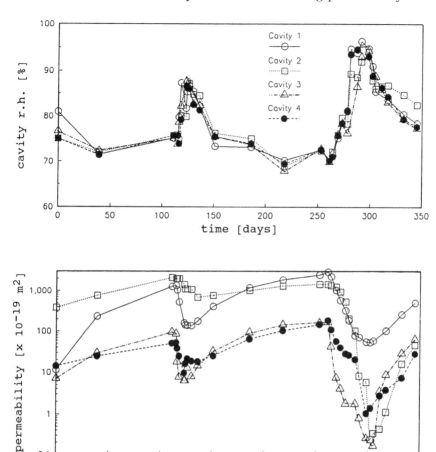

Fig. 4.37. Variations of relative humidity and permeability due to drying and wetting [4.56].

the formation of microcracks. Day *et al.* [4.35] showed that the method of drying significantly affected the measured value of oxygen permeability: oven drying at 105°C gave much higher values than those obtained by solvent exchange drying or direct drying at 20°C (Fig. 4.41).

Hudd [4.62] showed that the rate of air permeation increased with drying for concretes made with a range of water/cement ratios (Fig. 4.42). However, the effect of water/cement ratio appeared to be dependent upon the degree of saturation. Reinhardt [4.63] stated that air permeation was less affected by concrete moisture conditions if high (1 MPa) air pressures were used, but no experimental evidence was given to support this claim. Perraton *et al.* [4.64] observed that although oven drying caused

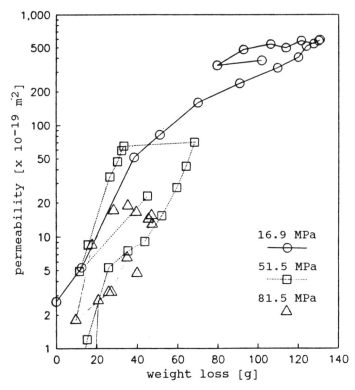

Fig. 4.38. Permeability versus weight loss of concrete cubes, for drying and wetting [4.57].

higher values of permeability than air drying the ranking of results for various dosages of silica fume was not much affected.

Thomas *et al.* [4.66] found that drying at 40% relative humidity gave permeability values about 15% higher than those for drying at 65% relative humidity.

Locher and Ludwig [4.67] reported that oxygen diffusion increased significantly with the severity and duration of drying (Fig. 4.43). Their results suggest that, in addition to drying, elevated temperatures cause detrimental change of pore structure. Hurling [4.68] found that drying for 3 months in the relative humidity range 100–65% greatly increased the oxygen diffusion coefficient, particularly when the relative humidity dropped below 85% (Fig. 4.44).

The results from many of the investigations of permeability and diffusion can be summarized and compared by treating the value corresponding to laboratory drying at about 60% relative humidity as unity: the relative values are shown in Table 4.31 and Fig. 4.45. Large increases of perme-

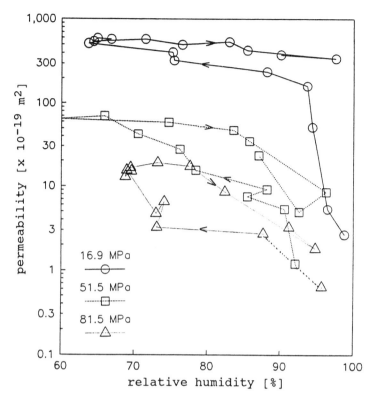

Fig. 4.39. Permeability versus cavity relative humidity in concrete cubes, for drying and wetting [4.58].

ability were generally observed for drying from saturation down to about 75% relative humidity but further drying to about 40% relative humidity had little effect.

These results are consistent with the idea that gas permeation is through emptied, capillary pores. The increase of permeability due to complete drying is probably associated with changes of pore structure and possibly cracking, rather than with emptying of pores.

Figures 4.35, 4.38, 4.39 and 4.42 suggest that where concretes have different porosities gas permeability is related more closely to the empty porosity (i.e. moisture loss) than to relative humidity. However, relative humidity can be readily measured in parallel with air permeability [4.53–4.59] whereas in-situ measurement of the empty porosity is uncertain. In the laboratory the empty porosity is easily monitored by weighing test specimens.

Table 4.31. Effect of moisture upon gas permeation (relative to value for about 60% **RH**)

Reference	Gas	Relative permeability (% RH)	Comments
4.53, 4.54	Air	2.3(35), 1.0(60, 0.02(85)	Dried to equilibrium, 0.4 to 0.6 w/c ratios
4.55	Air	1.0(60), 0.82(70), 0.60(80), 0.31(90), 0.18(94)	Reciprocal of Figg permeation time used.
4.56	Air	1.0(60), 0.80(70), 0.40(80), 0.18(90) 1.0(60), 0.77(70), 0.66(80), 0.58(90) 1.0(60), 0.80(70), 0.58(80), 0.45(90)	OPC concrete, not equilibrium values PFA concrete, not equilibrium values Slag concrete, not equilibrium values
4.57, 4.59	Air	1.0(60), 0.79(70), 0.55(80), 0.39(90) 0.06(95) 1.0(60), 0.72(70), 0.23(80), 0.11(90), 0.02(95) 1.0(60), 0.70(70), 0.30(80), 0.10(90), 0.05(95)	17 MPa concrete, not equilibrium values, drying* 52 MPa concrete, not equilibrium values, drying* 82 MPa concrete, not equilibrium values, drying*
4.58	Air	1.0(45), 1.0(65)	In situ, sheltered concrete column.
4.60	Air	12.8(Oven, 105C), 1.0(40)	Reciprocal of Figg permeation time used.
4.61 4.62	Air Air	8.3(Oven, 105C), 1.0(50% Sat.) 7.5(Oven, 50C), 1.0(70)	Degree of saturation used, % RH not given. Dried 3 to 40 weeks.
4.62	Air	2.8(20), 1.38(40), 1.0(50), 0.68(60), 0.20(80), 0.09(90)	Degree of saturation used, % RH not given.
4.64 4.66 4.68	Air Oxygen Oxygen	3.2(Oven, 105C), 1.0(50) 1.15(40), 1.0(65) 1.0(65), 0.58(75), 0.02(85), 0.02(96)	Concretes with various silica fume dosages Concretes with various curing periods and PFA contents Diffusion data

*Permeability values for wetting were higher.

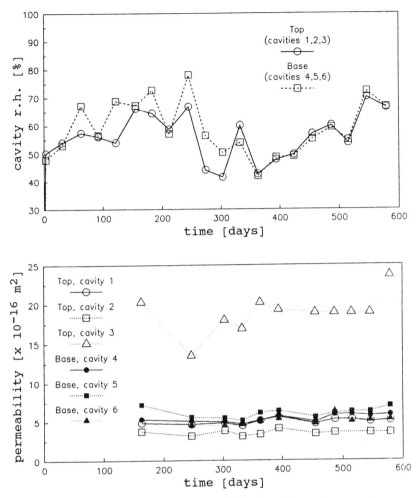

Fig. 4.40. Relative humidities and permeabilities in a sheltered, externally exposed concrete column [4.58].

4.3.3   EFFECT OF MOISTURE UPON WATER ABSORPTION

Hall, in his review of water absorption in concrete [4.43], emphasized the importance of having a uniform moisture condition at the start of a test. If this is not achieved then a square root time function for absorption will not be observed. In a theoretical analysis Hall suggested that the sorptivity S_i measured on a sample with an initial uniform degree of saturation M_l is given by $S_i = S_o (1 - 1.08M_l)^{0.5}$ where S_o is the sorptivity measured on a dry sample.

Millard [4.61] reported initial surface absorption (ISAT) data for a

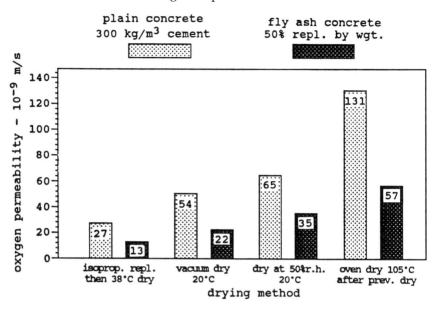

Fig. 4.41. Effect of drying method upon oxygen permeability [4.35].

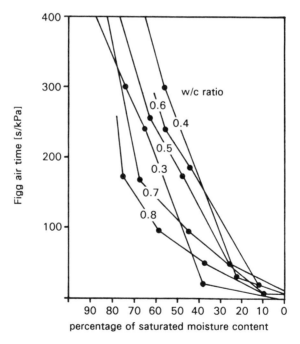

Fig. 4.42. Figg test permeation time versus moisture content for concretes of different water/cement ratios [4.62].

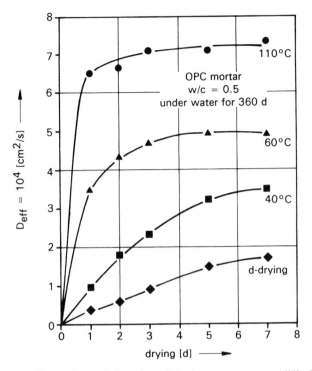

Fig. 4.43. Effect of severity and duration of drying upon oxygen diffusion coefficients [4.67].

concrete dried to different levels of saturation and tested at different temperatures (Fig. 4.46). It is evident that absorption increases with drying throughout the moisture content range and at each temperature. The results of Hudd [4.62] in Fig. 4.47 are similar to those of Millard but they are more detailed in the low moisture content range. For water/cement ratios 0.3–0.6 a maximum absorption is observed not for dry concrete, but for moisture contents between 10 and 30%. However, the maximum absorption for water/cement ratios of 0.7 and 0.8 occurs at zero moisture content. These results do not conform very closely to the equation describing the effect of moisture content proposed by Hall [4.43]. Figure 4.47 does not exhibit the expected trend of continuously increasing absorption with an increasing water/cement ratio.

Dhir et al. [4.69] also found that the severity and duration of drying affected the initial surface absorption (Fig. 4.48). It was concluded that oven drying at 105°C was a suitable conditioning method for laboratory testing; the more realistic air drying would be inconveniently slow.

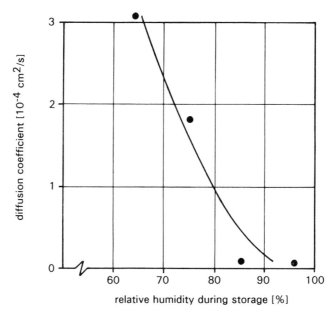

Fig. 4.44. Influence of storage humidity upon oxygen diffusion coefficients [4.68].

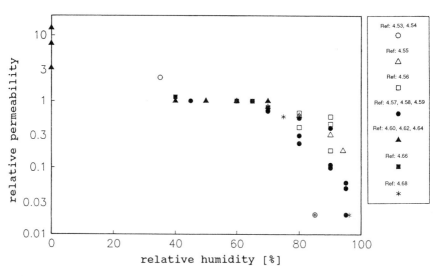

Fig. 4.45. Effect of relative humidity upon permeation data from different investigators (permeability values normalized by dividing by the value corresponding to 60% RH).

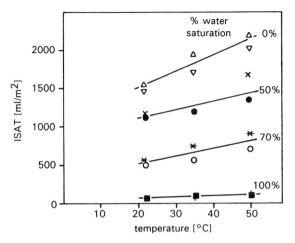

Fig. 4.46. Effects of moisture content and temperature upon initial surface absorption [4.61].

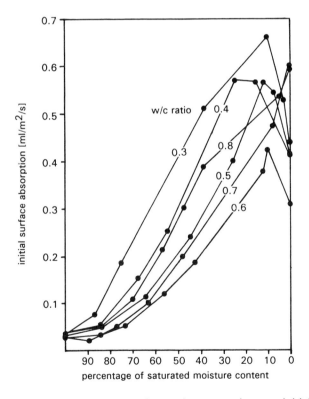

Fig. 4.47. Effects of moisture content and water/cement ratio upon initial surface absorption [4.62].

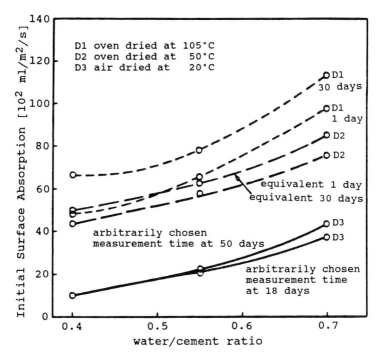

Fig. 4.48. Effect of drying method upon initial surface absorption of concretes with different water/cement ratios [4.69].

(a) Effects of temperature

The effect of temperature upon the viscosity of gases such as nitrogen and oxygen is small in the range 0–50°C. This is consistent with the small effect of test temperature upon Figg air permeation times, reported by Millard [4.61] (Fig. 4.49). The effect of temperature was not greatly influenced by moisture content.

Millard [4.61] found that the effect of temperature upon initial surface absorption shown in Fig. 4.46 could be predicted on the basis of known variations of water viscosity and surface tension with temperature. Figure 4.50 shows the temperature correction function recommended by Millard together with some averaged, experimental data.

(b) Effect of carbonation

Carbonation often reduces the capillary porosity of the cement paste matrix in cover concrete [4.69–4.74]. However, where large amounts of calcium silicate hydrate are formed (by replacing large amounts of Portland cement with fly ash or slag), carbonation can increase the capil-

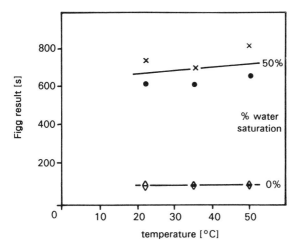

Fig. 4.49. Effect of temperature upon air permeation times; Figg test [4.61].

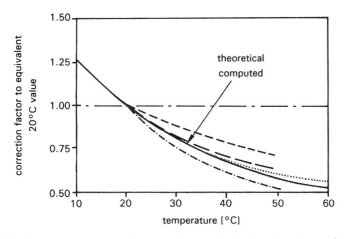

Fig. 4.50. Temperature correction function for initial surface absorption; theory and experiment [4.61].

lary porosity [4.73]. The permeability of cover concrete would be expected to change with carbonation in parallel to the changes of capillary porosity. Kropp [4.71] reported a reduced rate of water vapour diffusion in carbonated mortar. For an OPC mortar with $w/c = 0.5$, results are given in Fig. 4.51.

Bier [4.73] investigated the effect of carbonation on the pore structure of cementitious systems that were manufactured with different types of cement. It was found that for cements with a high clinker content, carbonation led to a considerable reduction of the capillary pore volume, whereas

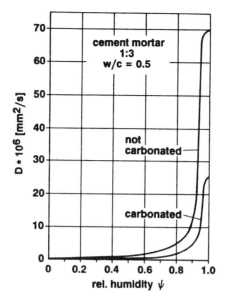

Fig. 4.51. Diffusion coefficient for water vapour through cement mortar [4.71].

for cements with a slag content exceeding approximately 50% by mass, a coarser capillary pore system was formed during carbonation. This was attributed to a considerable decomposition of the CSH gel in these cement pastes. The changes in the pore structure were reflected in the permeability of the different materials for water. Whereas systems containing OPC showed a decrease in permeability upon carbonation, PBFSC with slag contents exhibited a drastic increase in permeability after carbonation (cf. Fig. 4.52).

These results demonstrate that permeation of chemically reactive media through concrete may alter its permeability significantly, thus also controlling the resistance of concrete against other types of external attack. If the durability of concrete is expressed in terms of permeability these effects have to be taken into account.

Patel *et al.* [4.72] found that measured reductions of capillary porosity due to carbonation were associated with reduced rates of counter-diffusion, as pore water was exchanged with methanol. Martin [4.75] reported that nitrogen permeability measurements were repeatable whereas carbon dioxide permeability measurements were not; the observed reduction was attributed to the carbonation reaction. The changes of pore structure with carbonation are of practical significance for in-situ permeability testing of concrete and interpretation of results: measured values of permeability could exhibit a decrease over a long period of time, and such a decrease could be more pronounced in concretes made with a high water/cement

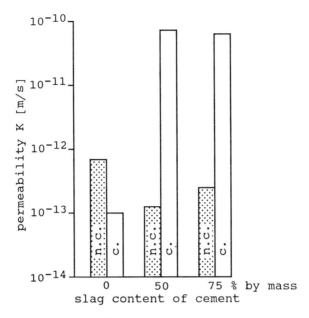

Fig. 4.52. Water permeability of different concretes before and after carbonation.

ratio. Thus it is necessary to make some allowance for carbonation if the condition or quality of the concrete at some earlier age is to be assessed from in situ permeability data.

4.3.4 CONCLUSIONS

The review of published data on the influence of environmental parameters upon concrete permeability suggests the following

1. Gas permeability is significantly increased as the capillary pores in the concrete are emptied; the channels available for gas flow increase in volume and connectivity. A further increase of permeability can be observed with severe drying, depending upon the temperature and duration of treatment. This is probably due to changes of microstructure in the cement paste matrix.
2. Water absorption is significantly increased as the capillary pores in the concrete are emptied, owing to the dependence of the hydraulic diffusivity upon moisture content. Severe drying causes a further increase of water absorption depending upon the temperature and duration of treatment.
3. The effect of test temperature upon gas permeation is small but water absorption is significantly affected. The increase in water absorption

with temperature is predictable in terms of changes in the viscosity and surface tension of water.

4. Carbonation can significantly affect permeability of cover concrete because the reaction modifies the pore structure of the cement paste matrix. Carbonation normally reduces permeability but the opposite effect may be observed where large amounts of OPC are replaced by mineral additions.

5. Environmental factors (relative humidity, temperature and atmospheric carbon dioxide) can significantly affect in-situ values of gas permeability and water absorption. Supplementary measurements relating to moisture conditions in the concrete and the depth of carbonation are desirable when interpreting in-situ permeability data.

6. Laboratory tests for gas permeability and water absorption should rigorously standardize the moisture conditions of the concrete. Carbonation should be avoided where possible; otherwise the depth of carbonation should be measured. Severe drying does not realistically simulate practical conditions and can alter the pore structure of the concrete: this matter requires consideration when it is necessary to accelerate the drying of laboratory test specimens.

4.4 References

4.1. Mehta, P.K. (1986) *Concrete: Structure, Properties and Materials*, Prentice-Hall.

4.2. US Bureau of Reclamation (1975) *Concrete Manual*, 8th edition, p. 37.

4.3. Adapted by Mehta from *Beton-bogen*, Aaborg, Denmark.

4.4. Maso, J.C. (1980) La liaison entre les granulats et la pâte de ciment hydraté, in *Proc. 7th International Congress on Chemistry of Cement*, Rapport principal, Vol. 1, Editions Septima, Paris.

4.5. Dhir, R.K., Hewlett, P.C. and Chan Y.N. (1987) Near surface characteristics of concrete: intrinsic permeability. *Magazine of Concrete Research*, Vol. 41, No. 147, June, pp. 87–97.

4.6. Nyame, B.K. (1985) Permeability of normal and lightweight mortars. *Magazine of Concrete Research*, Vol. 37, No. 130, pp. 44–8.

4.7. Watson, A.J. and Oyeka, C.C. (1981) Oil permeability of hardened cement pastes and concrete. *Magazine of Concrete Research*, Vol. 33, No. 115, pp. 85–95.

4.8. Powers, T.C. (1958) Structures and physical properties of hardened Portland cement pastes. *J. American Ceramic Society*, Vol. 41, No. 1, pp. 1–6.

4.9. Costa, U. and Massazza, F. (1958) ASMI-CNR-RILEM, in *2nd Int. Conf. Engineering Materials*, Bologna, pp. 8.

4.10. Hughes, D.C. (1985) Pore structure and permeability of hardened cement paste. *Magazine of Concrete Research*, Vol. 37, No. 133, pp. 227–33.

4.11. Massazza, F. (1988) Microscopic structure and concrete durability, in *Cracking of Concrete and Durability of Construction*, European Conference, AFREM – CCE, St-Rémy-lès-Chevreuses, pp. 5–15.

4.12. Powers, T.C. (1979) The specific surface area of hydrated cement obtained from permeability data. *Materials and Structures/Materiaux et Constructions*, Vol. 12, No. 69, pp. 159–68.

4.13. Huang, C.L.D., Siang, H.H. and Best, C.H. Heat and moisture transfer in concrete slabs. *Int. J. Heat and Mass Transfer,* Vol. 22, No. 2, pp. 257–66.

4.14. Powers, T.C., Copeland, L.E., Hayes J.C. and Mann, H.M. (1954) Permeability of Portland cement paste. *ACI Journal, Proceedings*, Vol. 51, No. 3, November, pp. 285–98.

4.15. Lawrence, C.D. (1985) Water permeability of concrete. Concrete Society Materials Research Seminar, Serviceability of Concrete, Slough, July.

4.16. Gräf, H. and Grube, H. (1986) Einfluß der Zusammensetzung und der Nachbehandlung des Betons auf seine Gasdurchlässigkeit, *Beton*, Vol. 36, No. 11, S. pp. 426–9 and No. 12, S. pp. 473–6.

4.17. Uchikawa, H. (1986) Effect of blending components on hydration and structure formation, in *Proc. 8th International Congress on the Chemistry of Cements*, Rio de Janeiro, September, Vol. 1, pp. 250–8.

4.18. Powers, T.C., Copeland, L.E. and Mann, H.M. (1959) Capillary continuity or discontinuity in cement pastes. *Journal of PCA Research and Development Laboratories*, Vol. 1, No. 2, May, pp. 38–48.

4.19. Bentz, D.P. and Garboczi, E.J. (1991) Percolation of phases in a three-dimensional cement paste microstructural model. *Cement and Concrete Research*, Vol. 21, pp. 325–44.

4.20. Parrott, L.J. (1985) Effect of changes in UK cements upon strength and recommended curing times. *Concrete*, Vol. 19, September, pp. 22–4.

4.21. Corish, A.T. and Jackson, P.J. (1982) Cement properties; past and present. *Concrete*, Vol. 16, July, pp. 16–18.

4.22. Nyame, B.K. and Illston, J.M. (1981) Relationships between permeability and pore structure of hardened cement paste. *Magazine of Concrete Research*, Vol. 33, No. 116, pp. 139–46.

4.23. Schönlin, K. and Hilsdorf, H.K. (1987) Evaluation of effectiveness of curing of concrete structures, in *Concrete Durability*, Katherine and Bryant Mather International Conference, ACI-SP100, pp. 207–26.

4.24. Grube, H. and Lawrence, C.D. (1984) Permeability of concrete to oxygen, in *Proc. of RILEM Seminar on Durability of Concrete Structures under Normal Outdoor Exposure*, Hanover, March, pp. 68–79.

4.25. Marsh, B.K. Relationship between engineering properties and microstructural characteristics of hardened pastes containing pfa as a partial cement replacement. PhD Thesis, Hatfield Polytechnic and Cement and Concrete Association.

4.26. Goto, S. and Roy, D.M. (1981) The effect of *w/c* ratio and curing temperature on the permeability of hardened cement pastes. *Cement and Concrete Research*, Vol. 11, pp. 575–9.

4.27. Häkkinen, T. Durability of alkali-activated slag concrete.

4.28. Kasai, Y., Matsui, I. amd Aoki, T. (1986) Long term changes of air permeability by rapid test. *Trans. Japanese Concrete Institute*, Vol. 8, pp. 145–52.

4.29. Figg, J.W. (1973) Methods of measuring the air and water permeability of concrete. *Magazine of Concrete Research*, Vol. 25, No. 85, December, pp. 213–19.

4.30. Manmohan, D. and Mehta, P.K. (1981) Influence of pozzolanic, slag, and chemical admixtures on pore size distribution and permeability of hardened cement pastes. *Cement and Concrete Research*, Vol. 3, No. 1, pp. 63–7.

4.31. Bakker, R.F.M. (1981) in *Proc. 5th. Int. Conf. Alkali-Aggregate Reaction in Concrete*, Cape Town, p. 7.

4.32. Hooton, R.D. (1986) *Blended Cement*, ASTM STP 897, American Society for Testing and Materials, pp. 128–43.

4.33. Marsh, B.K., Day, R.L. and Bonner, D.G. (1985) Pore structure characteristics affecting the permeability of cement paste containing fly ash. *Cement and Concrete Research*, Vol. 15, No. 6, pp. 1027–38.

4.34. Feldman, R.F. (1984) *J. American Ceramic Society*, Vol. 67, No. 1, pp. 30–4.

4.35. Day, R.L., Joshi, R.C., Langan, B.W. and Ward, M.A. (1985) Measurement of the permeability of concretes containing fly ash, in *Proc. 7th Int. Ash Utilization Symp. and Exposition*. Orlando, Vol. II, pp. 811–21.

4.36. Massazza, F. (1987) The role of the additions to cement in the concrete durability. *Il Cemento*, No. 4, pp. 359–82.

4.37. Gräf, H. and Grube, H. (1984) Oxygen permeability of concrete, in *Proc. of RILEM Seminar on Durability of Concrete Structures under Normal Outdoor Exposure*. Hanover, March, pp. 80–9.

4.38. Pomeroy, D. (1987) Concrete durability: from basic research to practical reality, in *Concrete Durability*, Katherine and Bryant Mather International Conference, ACI-SP100, pp. 111–30.

4.39. Schönlin, K. and Hilsdorf, H.K. (1989) The potential durability of concrete, ERMCO 89, The Norway to concrete, Stavanger, Oslo, Fabeko, 7 9 June.

4.40. Kettenacker, L, (1930) Über die Feuchtigkeit von Mauern, *Gesundheits-Ingenieur*, S. 721 ff.

4.41. Haque, M.N. (1990) Some concretes need 7 days initial curing. *Concrete International*, February.

4.42. Dhir, R.K., Hewlett, P.C. and Chan, Y.N. (1987) Near-surface characteristics of concrete: assessment and development of in situ test methods. *Magazine of Concrete Research*, Vol. 39, No. 141, December, pp. 183–95.

4.43. Hall, C. (1989) Water sorptivity of mortars and concretes: a review. *Magazine of Concrete Research*, Vol. 41, No. 147, June.

4.44. Kelham, S. (1988) A water absorption test for concrete. *Magazine of Concrete Research*, Vol. 40, No. 143, June, pp. 106–10.

4.45. Parrott. L.J. (1992) Water absorption in cover concrete. *Materials and Structures*, Vol. 25, No. 149, June, pp. 284–92.

4.46. Lydon, F.D. and Odaallah, M.Al. (1988) Discussion on paper published in *Magazine of Concrete Research*. Vol. 39, No. 141, December 1987, by R.K. Dhir, P.C. Hewlett and Y.N. Chan: Near-surface characteristics of concrete: assessment and development of in situ test methods. *Magazine of Concrete Research*, Vol. 40, No. 145, December.

4.47. Gowripalan, N., Cabrera, J.G. Cusens, A.R. and Wainwright, P.J. (1990) Effect of curing on durability. *Concrete International*, February.

4.48. Dhir, R.K. and Byers, E.A. (1991) PFA concrete: near surface absorption properties. *Magazine of Concrete Research*, Vol. 43, No. 157, December.

4.49. Marchese, B. and D'Amore, F. (1990) Discussion on paper published in *Magazine of Concrete Research*, Vol. 41, No. 147, June 1989, by C. Hall:

Water sorptivity of mortars and concretes: a review. *Magazine of Concrete Research*, Vol. 42, No. 151, June.

4.50. Dhir, R.K. and Yap, A.W.F. (1984) Superplasticized flowing concrete: durability properties. *Magazine of Concrete Research*, Vol. 36, No. 127, June.

4.51. Balayssac, J.P. (1992) Relations entre performances mécaniques microstructure et durabilité des bétons, Thèse de l'Institut National des Sciences Appliquées de Toulouse, June.

4.52. Thomas, M.D.A., Osborne, G.J., Matthews, J.D. and Cripwell, J.B. (1990) A comparison of the properties of OPC, PFA and ggbs concretes in reinforced concrete tank walls of slender section. *Magazine of Concrete Research*, Vol. 42, No. 152, September, pp. 127–34.

4.53. Ujike, I. and Nagataki, S. (1988) A study on the quantitative evaluation of air permeability of concrete. *Proc. Japanese Society Civil Engineers*, Vol. 9, No. 396, pp. 79–87.

4.54. Nagataki, S., Ujike, I. and Konishi, N. (1986) Influence of moisture content on air permeability of concrete, in *Review of 40th Meeting of Cement Association of Japan*, Tokyo, pp. 158–61.

4.55. Jonis, J. and Molin, C. (1988) *Measuring of air permeability of concrete*, Swedish National Testing Institute Report 1988: 4.

4.56. Chen Zhang Hong and Parrott, L.J. (1989) *Air permeability of cover concrete and the effect of curing*, British Cement Association Report C/5, October, 25 pp.

4.57. Parrott, L.J. and Chen Zhang Hong. (1990) Some factors influencing air permeation measurements in cover concrete. *Materials and Structures*, Vol. 24, No. 144, November, pp. 403–8.

4.58. Parrott, L.J. (1990) Assessing carbonation in concrete structures, in *Durability of Building Materials and Components*, Proceedings of Fifth International Conference, Brighton, UK, Baker, J.M. *et al.* (Eds), E & FN Spon, London, pp. 575–86.

4.59. Parrott, L.J. (1990) Unpublished data on relationships between air permeability and relative humidity in cover concrete, January.

4.60. Pihlajavaara, S.E. and Paroll, H. (1975) On the correlation between permeability properties and strength of concrete. *Cement and Concrete Research*, Vol. 5, No. 4, pp. 321–8.

4.61. Millard, S. (1989) Effects of temperature and moisture upon concrete permeability and resistivity measurements, Workshop on in situ permeability, Loughborough, December, 9 pp.

4.62. Hudd, R. (1989) Effect of moisture content on in situ permeability readings, Workshop on in situ permeability, Loughborough, December, 6 pp.

4.63. Reinhardt, H.W. and Mijnsbergen, J.P.G. (1989) In-situ measurement of permeability of concrete cover by overpressure, in *The Life of Structures: Physical Testing*, Proceedings of International Seminar, Brighton, April, Butterworths, London, pp. 243–54.

4.64. Perraton, D., Aitcin, P.C. and Vezina, D. (1988) Permeabilities of silica fume concrete, in *Permeability of Concrete*, ACI Special Publication SP-108, pp. 63–84.

4.65. Nagataki, S. and Ujike, I. (1980) Effect of heating condition on air permeability of concrete at elevated temperature. *Transactions of the Japanese Concrete Institute*, Vol. 10, pp. 147–54.

4.66. Thomas, M.D.A., Matthews, J.D. and Haynes, C.A. (1989) The effect of curing on the strength and permeability of PFA concrete, in *Flyash, silica fume, slag and natural pozzolans in concrete, Proceedings 3rd International Conference*, V.M. Malhotra (Ed.), Trondheim, ACI SP-114, Vol. 1, pp. 191–217.

4.67. Locher, C. and Ludwig, U. (1987) Measuring oxygen diffusion to evaluate the open porosity of mortar and concrete. *Betonwerk + Fertigteil-Technik*, Vol. 3, pp. 177–82.

4.68. Hurling, H. (1984) Oxygen permeability of concrete, in *Proc. RILEM Seminar on Durability of Concrete Structures under Normal Outdoor Exposure*, Hanover, March, pp. 91–101.

4.69. Dhir, R.K, Hewlett, P.C. and Chan, Y. (1987) Discussion of their Paper in *Magazine of Concrete Research*, Vol. 39, No. 141, December 1987. *Magazine of Concrete Research*, Vol. 40, No. 145, pp. 234–44.

4.70. Parrott, L.J. (1987) *A review of carbonation in reinforced concrete*, British Cement Association Report C/1-0987, 121 pp.

4.71. Kropp, J. (1983) Influence of carbonation on the structure of hardened cement paste and water transport, in *Proc. International Colloquium*, Esslingen, pp. 153–7.

4.72. Patel, R., Parrott, L.J., Martin, J. and Killoh, D. (1985) Gradients of microstructure and diffusion properties in cement paste caused by drying. *Cement & Concrete Research*, Vol. 15, No. 2, pp. 343–56.

4.73. Bier, T. (1987) Influence of type of cement and curing on carbonation progress and pore structure of hydrated cement pastes, in MRS Symp. Proc. Vol. 85, *Microstructural Development during Hydration of Cement*, pp. 123–34.

4.74. Parrott, L.J. (1987) Measurement and modelling of porosity in drying and cement paste, in *Microstructural Development during Hydration of Cement*, MRS Symp. Proc. Vol. 85, pp. 91–104.

4.75. Martin, G. (1986) A method for determining the relative permeability of concrete using gas. *Magazine of Concrete Research*, Vol. 38, No. 135, June, pp. 90–4.

5 Relations between transport characteristics and durability

Jörg Kropp

5.1 Introduction

Concrete as a porous material may interact with its environment: i.e. compounds of concrete may be leached out, thus weakening the structure of the concrete, or material from the environment may penetrate into the concrete, thereby causing chemical or physical interactions with concrete compounds. These interactions can cause degradation processes of the concrete itself or of the reinforcement embedded in the concrete.

Depending on the type of attack the transport mechanism involved in the degradation process consists in the permeability of water or aqueous solutions or absorption by capillary action, respectively, the diffusion of gaseous compounds through the open pore system or ionic diffusion of dissolved compounds into or out of concrete. For concrete in service a combined action of various media may prevail and mixed modes of transport processes occur.

It has been well established that the perviousness of concrete plays an important role in the control of concrete durability [5.1–5.3]. Accordingly, testing of transport parameters for concrete, such as permeability, diffusivity or absorption behaviour in most cases has been done on the basis of durability considerations.

Since the early work of Levitt [5.4] and Figg [5.5] on the development of various test methods for concrete transport parameters, special interest is growing in the assessment of concrete durability by means of its transport characteristics – either by laboratory testing of drilled cores, measurements on separately cast companion specimens, or with the help of new test methods available also by on-site testing of the structure concerned [5.6]. More recent developments consider a classification of concrete according to its durability [5.7]. In such a concept the transport characteristics could serve as compliance criteria.

Basics of these approaches have already entered national or international standards and recommendations: e.g. stating limiting values for

Performance Criteria for Concrete Durability, edited by J. Kropp and H.K. Hilsdorf. Published in 1995 by E & FN Spon, London. ISBN 0 419 19880 6.

the depth of water penetration or absorption rates [5.8–5.12] for concrete that may be subject to chemical attack or frost action. It has been pointed out, however, that a major shortcoming of this approach is still a lack of data on the correlation between any of the available transport character-istics and the resistance of concrete to physical and chemical attack [5.1]. Therefore, those requirements or recommendations set forward already in standards and codes are based on empirical considerations rather than on quantitative analyses.

In the following an attempt is made to collect published data on corre-sponding correlations. However, very often such correlations are discussed only on a qualitative basis.

5.2 Carbonation

The alkaline compounds of the hydrated cement paste matrix react with penetrating carbon dioxide to form carbonates, thereby liberating water and/or metal oxides depending on the hydration product involved. Also, the major constituent of hydrated cement paste, the CSH gel, is subject to carbonation. The gel is decomposed into calcium carbonate and an amorphous silica gel with a porous structure [5.13].

Although carbonation may not be harmful for the concrete itself, these reactions are accompanied by a drop of the pH of the concrete pore water. Thus the corrosion protection of reinforcement by passivation in an alka-line environment is lost. For reinforced concrete structures the onset of reinforcement corrosion is often considered as determining the lifetime of the structure [5.14].

Carbonation rate is controlled by the ingress of carbon dioxide into the concrete pore system by diffusion, with a concentration gradient of CO_2 acting as the driving force. Thereby, the carbon dioxide always has to pass through concrete sections that have already undergone carbonation.

Therefore, Fick's law of diffusion has been applied to describe the depth of carbonation as a function of time [5.15]. Based on Fick's first law of dif-fusion, the amount m of CO_2 diffusing through a concrete layer is given by

$$m = -DA\frac{c_1 - c_2}{x}t$$

$$(5.1)$$

where m = mass of carbon dioxide (g)
   D = diffusion coefficient for carbon dioxide through carbon-ated concrete (m^2/s)
   A = penetrated area (m^2)
   c_1 = external concentration of carbon dioxide (g/m^3)
   c_2 = concentration of carbon dioxide at the carbonation front (g/m^3)

t = time (s)
x = thickness of penetrated concrete layer (m)

At the carbonation front the penetrated CO_2 reacts with the alkaline compounds available in a given concentration. For the carbonation of these alkaline compounds contained in a unit volume of concrete an amount of CO_2 a (g/m³) is required, and

$$m = a\,A\,dx \qquad (5.2)$$

gives the mass of CO_2 required to increase the depth of carbonation by an increment dx.

Inserting eq. (5.2), eq. (5.1) may be rewritten as

$$a\,A\,dx = -D\,A\frac{c_1 - c_2}{x}t \qquad (5.3)$$

or

$$x\,dx = -\frac{D}{a}\left(c_1 - c_2\right)t \qquad (5.4)$$

Integration of eq. (5.4) finally yields

$$x^2 = \frac{2D}{a}\left(c_1 - c_2\right)t \qquad (5.5)$$

If all constant parameters of eq. (5.5) are combined into one single constant C, eq. (5.5) will result in the well-known equation

$$x = C\sqrt{t} \qquad (5.6)$$

where x = depth of carbonation at time t (m)
       t = time (s)
       C = constant

In the derivation of eqs (5.1)–(5.6) several simplifications have been made:

1. The diffusion coefficient for carbon dioxide through carbonated concrete has been taken as a constant material property. For a concrete structure in service, however, D may be a function of numerous variables. The diffusion coefficient may depend on the location due to changes in the pore structure of the hcp, e.g. due to curing effects. Furthermore, the moisture content is an important parameter. For a high moisture concentration the pore system is filled with water and the gas diffusion ceases [5.16–5.18].

Water-saturated concrete therefore does not undergo any significant carbonation. On the other hand, very dry concrete is very permeable for gas diffusion. Nevertheless, carbonation rate will be low because the reaction between CO_2 and the alkaline compounds requires some minimum moisture content for dissociation. Highest carbonation rates are found for cementitious materials that are in equilibrium with ambient air of approximately 50% RH [5.19, 5.20].

Concrete members exposed to natural weathering conditions will have varying moisture contents with time, and the moisture distribution across the concrete section may exhibit steep profiles. Therefore, the diffusion coefficient will vary with time t as well as with location x.

Carbon dioxide has to penetrate through carbonated material; therefore the diffusion coefficient D of the carbonated material controls the ingress of CO_2. This parameter may not be constant over time because ongoing carbonation reactions in the 'carbonated zone' will affect the pore structure and thus the diffusivity.

The effect of temperature on D may be twofold: the gas diffusion will be enhanced by elevated temperatures; additionally, the moisture content will be reduced thus providing more free pore space for gas diffusion.

2. The derivation implies that the depth of carbonation is given by a sharp reaction front separating non-carbonated material from carbonated material in which all alkaline compounds have been transformed. In reality a transition zone develops with respect to the formation of carbonates. In this zone also the **depth of carbonation** falls, which is usually determined by means of sprayed indicator solutions such as phenolphthalein, for example according to RILEM Technical Recommendation CPC 18, *Measurement of hardened concrete carbonation depth* [5.21]. This test, however, only distinguishes regions of high pore water alkalinity from areas of lower alkalinity depending on the indicator solution used. Carbon dioxide may have penetrated beyond the indicated depth of carbonation not yet leading to a drop of the pH; at the same time no complete carbonation of all alkaline compounds may have occurred in the neutralized zone with some carbonation reactions still going on.

3. The amount of carbon dioxide a (g/m^3) that is needed for carbonation of the alkaline compounds contained in a unit volume of cementitious material may not be constant. In the first place this parameter depends on the type of the cement, especially on its CaO content [5.22], the mix proportions of the concrete and the type as well as the amount of pozzolanic additions. Methods to calculate the parameter a from data on the cement composition, type and concentration of mineral additions and mix proportions of the concrete are given for example in [5.16, 5.17]. Furthermore, for a given concrete the parameter a depends

on the degree of hydration of the cement. Owing to curing effects the degree of hydration may exhibit a gradient from the inner sections of a concrete member towards the surface. Additionally, a depends on the CO_2 concentration. For example, under atmospheric CO_2 concentration $Mg(OH)_2$ does not undergo carbonation; however, transformation into carbonates occurs only for higher CO_2 concentration [5.22].

In an experimental investigation on carbonation [5.23] a higher degree of decomposition of the CSH gel was found for test conditions using a CO_2 content of the air of 2% by volume compared with companion specimens that were stored in natural air. Because CSH can be decomposed at natural CO_2 concentration [5.22] this difference is a time effect rather than a concentration effect.

The degree of carbonation is given by the amount of alkaline compounds transformed into carbonates in relation to the amount of substance that can be transformed into carbonates [5.22]. In laboratory experiments, as well as for outdoor exposure, values for the degree of carbonation between 0.4 and 0.8 [5.23–5.25] have been reported. This implies that the parameter a may not be computed for a given cement assuming a total carbonation of all alkaline compounds. The value of a is likely to be a function of time as well as of exposure conditions.

The simplifications mentioned above demonstrate that the application of Fick's law of diffusion using constant parameters is subject to many limitations. A variety of prediction models for the development of the depth of carbonation with time exist [5.13] based on this theory. But especially for higher concrete ages the depth of carbonation observed is often less than expected according to a $\sqrt{t}$ relation. Therefore, time functions $x = A\ t^n$ with $n < 1/2$ have been proposed [5.26]. In constant climates, e.g. laboratory experiments, satisfactory results may be obtained; however, correlation with the observed carbonation rate is often poor, especially for outdoor exposure with changing humidity conditions. Based on long-term observations of the carbonation rate of concretes subjected to natural weather conditions Wierig [5.27] concluded that a $\sqrt{t}$ approach is sufficiently accurate to describe the progress of carbonation as long as the concretes are sheltered from rain. Deviations from a $\sqrt{t}$ relation, however, occurred for concretes that could absorb rain water (see also Chapter 7).

Taking into consideration that high moisture concentrations in different depths from the concrete surface will reduce the CO_2 diffusion significantly, simultaneously allowing a counter-diffusion of alkalis from interior sections towards the carbonation front, Schießl [5.25] arrives at the following equation:

$$t = -\frac{a}{b}\left[x - x_{oo} \ln\left(1 - \frac{x}{x_{oo}}\right)\right]$$

(5.7)

where t = time (s)
       a = amount of carbon dioxide needed for carbonation of unit
             volume of concrete [5.9]
       b = factor describing the decrease of the diffusion coefficient
             for CO_2 as well as the CO_2 demand of the counter-
             diffusing alkalis
       x = depth of carbonation (mm)
       x_{oo} = final depth of carbonation (mm)

Another approach presented by Bakker [5.18] considers discrete periods
of wetting and drying of the near-surface concrete layers and their respective carbonation rates.

As a major implication Schießl's method as well as the approach of
Bakker result in a final depth of carbonation. This is not the case for any
method following simple t^n relations, e.g. a $\sqrt{t}$ approach.

Papadakis *et al.* [5.16, 5.17] have measured the diffusion coefficient
for CO_2 of concretes manufactured with different types of cement as a
function of the equilibrium moisture content. For their results as shown
in Fig. 5.1 an empirical relation was found between the diffusion coefficient
for carbon dioxide and the relative humidity.

Based on these results, the authors calculated the progress of carbonation for different, but constant, relative humidities, thereby applying
diffusion theory.

For a more realistic description of the progress of carbonation during
outdoor exposure it is necessary to consider a coupled transport process
in which the diffusion of CO_2 is correlated to the drying and re-wetting
of the concrete during changing moisture conditions. Corresponding
attempts have been made earlier by Houst and Wittmann [5.28], and the
results of Papadakis *et al.* [5.16, 5.17] could serve as valuable input in
further calculations.

In principle, eqs (5.1)–(5.6) could also be used to calculate the diffusion coefficient D for CO_2 from observed carbonation rates of a given
concrete. Considering the above-mentioned limitations this approach may
only result in a rough estimate for the diffusion coefficient.

Aside from numerous attempts to correlate the depth of carbonation
with concrete compressive strength [5.26], recent research work has established correlations with the permeability of concrete for gases as well as
with the absorption of water [5.29–5.35].

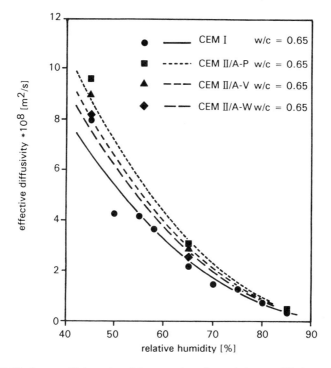

Fig. 5.1. Diffusion coefficient for CO_2 as a function of the equilibrium moisture content [5.16, 5.17].

5.2.1 GAS PERMEABILITY

The correlation between air permeability of concrete and its depth of carbonation after one year storage in a controlled atmosphere of 20°C and 65% RH with natural CO_2 content was studied in [5.29]. The different concrete mixes were manufactured with varying water cement ratios, different types of cement and varying amounts of fly ash additions or with inert filler (ground limestone) as well as air-entrained mixes made of Portland cement. After demoulding, the concrete samples were subjected to wet curing for periods varying between 1 and 7 days. Subsequent to curing the specimens were stored at 20 C and 65% RH up to an age of 56 days. Air permeability then was measured by means of the vacuum method described in section 9.4.

In Fig. 5.2 the correlation between the depth of carbonation of concrete and the air permeability is given for different types of cement used in manufacturing the concrete mixes.

Figure 5.3 gives corresponding results for concretes made with OPC and varying amounts of fly ash additions.

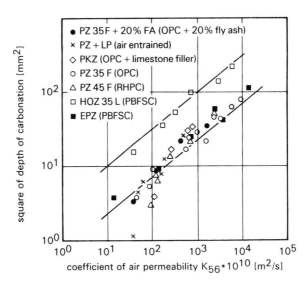

Fig. 5.2. Depth of carbonation after 1 year storage in normal air of 65% RH as a function of concrete air permeability at an age of 56 days [5.29].

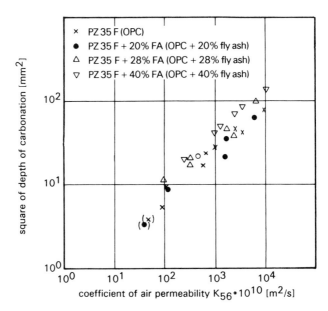

Fig. 5.3. Depth of carbonation for fly ash concretes after 1 year storage in normal air of 65% RH as a function of the concrete air permeability at an age of 56 days [5.29].

These results indicate that a linear relationship may be established if the square of the depth of carbonation is plotted against the logarithm of the air permeability.

This linear relationship has also been derived analytically by Hilsdorf [5.3]. Experimental results on oxygen permeability and oxygen diffusivity obtained by Lawrence [5.36] indicate that a linear relationship exists between the logarithms of the coefficient of diffusion and coefficient of permeability for oxygen. This relationship, which is shown in Chapter 3, Fig. 3.8, may be given as

$$\left(\frac{D}{D_0}\right) = \left(\frac{K}{K_0}\right)^n$$

(5.8)

where D = diffusion coefficient (m²/s)
K = coefficient of permeability (m²)
n = exponent
D_0 = constant
K_0 = constant

Solving eq. (5.8) for D and inserting the result into eq. (5.5) will yield

$$x^2 = \frac{2(c_1 - c_2)D_0}{a}\left(\frac{K}{K_0}\right)^n .t$$

(5.9)

Equation (5.9) now postulates a linear relationship between the square of the depth of carbonation, x^2, and the logarithm of concrete gas permeability, the exponent n representing the slope of the straight lines in Figs 5.2 and 5.3.

In Fig. 5.2, for a given permeability a higher rate of carbonation is shown for PBFS cements with a high slag content of 65% by mass as compared to cements with a high clinker content. This behaviour is caused by the coarsening of the pore structure of the matrix of PBFS cements upon carbonation, thus increasing the permeability. In the measurement of the gas permeability, however, the air flow occurred across the specimen thickness with only two thin layers of carbonation at the opposite ends of the concrete discs. If the permeability is tested on carbonated samples, however, the rate of carbonation may be represented by one linear relation independent of the type of cement [4.33]. These results are presented in Fig. 5.4. The permeability index used in this diagram has been determined on the concrete surface with the help of an on-site test method described in section 10.2 (Schönlin method). For all storage conditions the age of the concrete specimens was 56 days.

The effect of curing on oxygen permeability as well as on durability characteristics such as rate of carbonation has been investigated in [5.37],

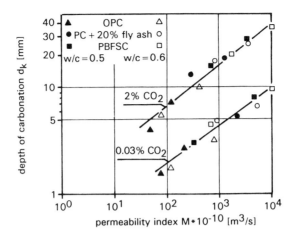

Fig. 5.4. Depth of carbonation of concrete after 56 days storage in air with 65% RH as a function of the surface permeability index [5.33].

using ordinary Portland cement, PBFSC and fly ash cement in manufacturing the test specimens. The following curing regimes have been tested:

A   1 day in mould, 1 day exposed to wind, 26 days at 65% RH
B   1 day in mould, 27 days at 65% RH
C   1 day in mould, 2 days wrapped in plastic sheet, 25 days at 65% RH
D   1 day in mould, 26 days wrapped in plastic sheet
E   1 day in mould, 6 days submerged in water, 21 days at 65% RH

In Fig. 5.5 the depth of carbonation is given after 3 years' exposure to natural air of 65% RH and outdoor climate, respectively, as a function of the oxygen permeability, which was measured at a concrete age of 35–56 days. Also, these results demonstrate that an increase in permeability favours a higher rate of carbonation. Again, for a given permeability, concretes made of PBFSC with a slag content of 36% by mass show higher depths of carbonation than PC or fly ash cement.

Dhir [5.38–5.41] has studied the transport characteristics of near-surface concrete layers by means of different test methods. One point of interest had been the correlation between the rate of carbonation and the air permeability of concrete. Aside from a laboratory method for the determination of the coefficient of air permeability a modified Figg test set-up suitable for on site testing was used to derive an air permeation index.

Four different groups of concrete test series were manufactured with water/cement ratios ranging from 0.4 to 0.7, different workabilities, varying maximum aggregate sizes and special additions such as fly ash, air-entraining agent or superplasticizer. The design strength of the mixes

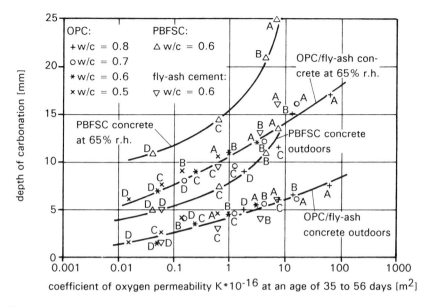

Fig. 5.5. Depth of carbonation after 3 years as a function of oxygen permeability [5.37].

was between 30 and 65 N/mm². After demoulding four different curing methods were chosen:

1. Water curing at 20°C for 28 days
2. Water curing for 6 days, 21 days in air of 55% RH at 20°C
3. Water curing for 3 days, 24 days in air of 55% RH at 20°C
4. 27 days air of 55% RH at 20°C.

After the curing period the test specimens for permeability measurements were oven dried at 105°C, whereas the specimens for carbonation were stored in a controlled atmosphere of 55% RH at 20°C for 14 days in order to achieve some pre-drying prior to carbonation.

For carbonation, accelerating conditions were chosen, i.e. a carbon dioxide content of the air of 4% by volume; relative humidity was controlled to 50% RH at 20°C. The progress of carbonation was observed by means of the phenolphthalein test up to 20 weeks of exposure to CO_2.

The test results obtained showed a very close correlation such that permeability measurements either in laboratory experiments or with on-site methods (modified Figg set-up) were recommended for predicting the progress of carbonation with time.

In Fig. 5.6 results are given for air permeability and depth of carbonation after a total of 20 weeks exposure to accelerating carbonation conditions.

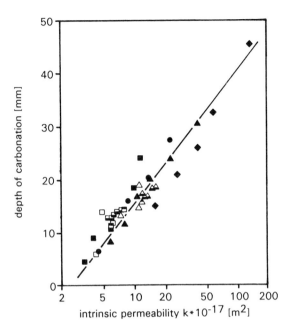

Fig. 5.6. Relationship between permeability k and depth of carbonation [5.40].

Plotting the coefficient of air permeability in a logarithmic scale, a linear regression line is obtained.

Comparing the depth of carbonation in accelerated tests with that obtained on companion specimens stored outside, but protected from rain, it was found that one week under accelerating conditions corresponded to 15 months of natural exposure.

From these results the following prediction model was derived:

$$D_t = \left(\frac{t}{20}\right)^\gamma \left(22.8 \ \log \ k \ - \ 6.9\right)$$

(5.10)

where D_t =   depth of carbonation (mm) after t years of normal exposure
       k =   coefficient of air permeability (m^2)
       γ =   0.5 for $w/c \leq 0.6$, 0.4 for $w/c > 0.6$

Figure 5.7 shows the relation between depth of carbonation and the Figg air permeation index. Also, for this test method a linear regression line was found if the depths of carbonation after 20 weeks of accelerating conditions were plotted against the logarithms of the Figg index. This correlation has also been used for a prediction model for carbonation depths:

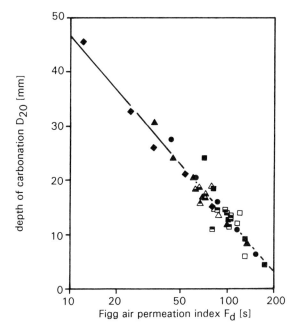

Fig. 5.7. Relation between depth of carbonation after 20 weeks of accelerated carbonation and Figg air permeation index [5.40].

$$D_t = \left(\frac{t}{20}\right)^{\gamma}\left(82 - 34\log F_d\right)$$

(5.11)

where F_d = Figg air permeation index (s) and others as given above.

In [5.30], field observations are given for the correlation between air permeability and depth of carbonation of structural concrete members (Fig. 5.8). The concrete members investigated had an age between 3 years and 38 years. For comparison the observed depth of carbonation was normalized for a concrete age of 25 years using a $\sqrt{t}$ approach. The gas permeability was measured by means of the Paulmann borehole method described in section 10.2. Although there is some scatter of the experimental readings the results indicate a marked increase of carbonation depth with increasing concrete permeability. Also, experiments published by Kikuchi *et al.* [5.42] support a correlation between carbonation rate and air permeability.

Based on experimental results, Osborne [5.34] also found good correlations between nitrogen permeability and depth of carbonation for concretes exposed to outdoor climate.

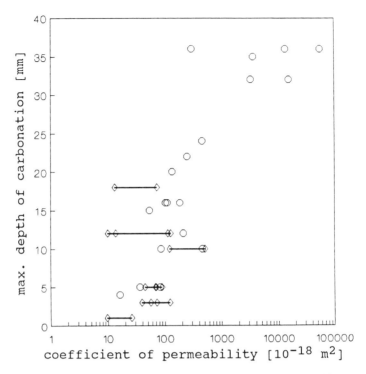

Fig. 5.8. Observed depth of carbonation and concrete air permeability for structural concrete members [5.30].

5.2.2 CAPILLARY SUCTION OF WATER

Further experimental investigations show that carbonation rate also correlates with the absorption of water due to capillary action. In Fig. 5.9 results obtained on different concrete mixes are presented [5.29]. In this diagram a second relation is valid for concrete mixes either air-entrained or manufactured with PBFSC with a slag content of 65% by mass. For PBFSC this behaviour may be explained by changes in the pore structure upon carbonation. Air-entrained concrete mixes of given permeability show less absorption of water because their capillaries may be interrupted by air voids, thus reducing the absorption rate.

In [5.35] the depth of carbonation of concrete cubes observed after 0.5 years and 1.5 years exposure to laboratory or outside conditions, respectively, is correlated to the absorption of water after a wetting period of 4 h. This criterion was preferred because no $\sqrt{t}$ relation was observed. In Fig. 5.10 the results for 1.5 years exposure are presented. In contrast to air permeability the absorption of water showed a poorer correlation with carbonation rate in experiments published by Kikuchi *et al.* [5.42].

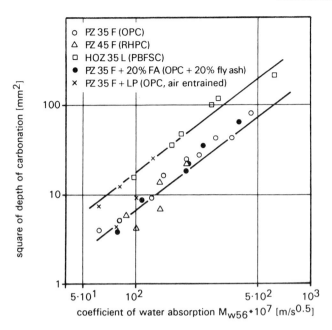

Fig. 5.9. Depth of carbonation after 1 year storage in normal air of 65% RH as a function of the absorption rate measured at an age of 56 days [5.29].

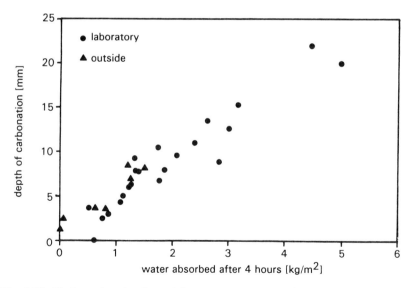

Fig. 5.10. Carbonation depths at 1.5 years versus water absorption after 1.5 years of exposure [5.35].

5.3 Sulphate attack

Sulphates carried into the inner sections of concrete by either ionic diffusion or capillary absorption of sulphate solutions may cause disruptive forces leading to cracking or scaling of the concrete. The corrosive action of sulphates, however, depends on the presence of reactive hydration products of the cement in sufficient concentration, e.g. calcium hydroxide, CAH and CASH respectively. Penetrating sulphates may react with these phases, forming gypsum or ettringite. These expansive phases then may disrupt the structure of the concrete [5.43].

Concrete of sufficiently low permeability is considered as non-susceptible to sulphate attack [5.12]. Furthermore, sulphate-resisting cements with low concentrations of sulphate-reactive compounds may not undergo expansive phase transformations even upon access of sulphates, and the transport capacity of corresponding concrete mixes may then not reflect its resistance to sulphate attack in a correct manner.

In experimental studies on high-strength concretes also the resistance against sulphate attack was investigated and correlated with physical properties of the concretes [5.44].

In Table 5.1 the mix proportions of 10 different mixes are given as well as their 28-day compressive strength. Ordinary Portland cement was used. Sulphate resistance was tested in a crystallization test according to RILEM Tentative Recommendation, Recommended tests to measure the deterioration of stone and to assess the effectiveness of treatment methods, Part V, 'Durability Tests', Test No. V.1 'Crystallization test by total immersion' [5.45].

In these investigations a close correlation was established between the maximum weight gain of the test samples during 120 soaking/drying cycles

Table 5.1 Concrete composition and compressive strength [5.44]

Concretes										
	B1	*B2*	*B3*	*B4*	*B5*	*B6*	*B7*	*B8*	*B9*	*B10*
Cement (kg/m^3)	500	500	450	450	450	450	450	450	450	450
Aggregates (kg/m^3)	1733	1828	1839	1815	1620	1616	1815	1801	1706	1643
$w/(c+a)$	0.36	0.24	0.25	0.24	0.42	0.42	0.24	0.25	0.36	0.40
Superplasticizer (l/m^3)	–	15	15	15	–	–	15	15	–	–
Silica fume A	–	–	105	–	76	–	–	–	–	–
Silica fume B	–	–	–	50	–	50	–	–	–	–
Fly ash	–	–	–	–	–	–	50	–	50	–
Pozzolana	–	–	–	–	–	–	–	50	–	50
Compressive strength	67.2	83.7	106.2	105.4	74.9	72.4	86.7	90.7	66.0	61.2

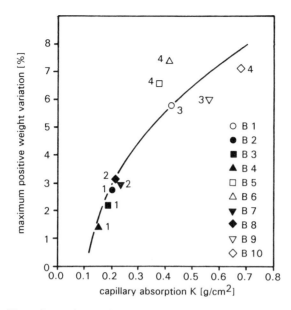

Fig. 5.11. Capillary absorption and weight change of concrete after sulphate attack [5.44].

and the capillary absorption of the concrete series, which had been tested according to RILEM Tentative Recommendation No. 11.2, Absorption of water by capillarity [5.46]. The maximum weight gain had been chosen as a criterion of the sulphate resistance because in most cases a marked increase in weight during the first 30 cycles was associated with considerable weight losses later on. In Fig. 5.11 the maximum weight gain is plotted against capillary absorption. In this diagram numbers are attributed to the individual symbols. These numbers describe a classification of the sulphate resistance based on a visual inspection of the concrete samples after completion of the test. This classification comprises four groups:

1. no trace of attack;
2. slight attack;
3. moderate attack;
4. severe attack.

This behaviour may be explained by the specific corrosion test chosen, in which oven-dried concrete samples were repeatedly soaked in sulphate solutions.

Furthermore, it had been attempted to correlate the sulphate resistance to the permeability of the concrete, which had been determined according to RILEM Tentative Recommendation CPC 13.2, Test for permeability of porous concrete [5.47]. A poor correlation has been reported, however.

Van Aardt observed an increasing resistance of concrete against sulphate attack for samples that exhibited a carbonated surface zone. This improvement was partially attributed to a restricted ingress of sulphates due to a densified surface zone upon carbonation; however, carbonation also transforms reactive phases to more stable compounds, which are less vulnerable to sulphate attack [5.48].

Much attention has been given to the effect of mineral additions on the sulphate resistance of concrete, e.g. fly ash or silica fume [5.49–5.52]. However, the observed effects have been discussed deeper on a phenomenological basis only. Accordingly, improvements in the sulphate resistance will result from an impermeable pore structure due to the finely dispersed hydration products of the pozzolanic additions, reducing the ingress of sulphate ions. Simultaneously, the chemical composition of the matrix is also changed such that the pozzolanic activity of the additions forms phases chemically more stable than the hydration products of Portland cement clinker [5.52, 5.53].

The performance of concrete containing fly ash may depend on the type of fly ash used, because the fly ash itself may incorporate reactive phases such as reactive aluminates or sulphates. A beneficial effect of a denser pore structure may then be offset [5.54].

Diffusion coefficients for sulphate ions into hcp are reported to be 10–100 times lower than those for chlorides, ranging from 2 to 30×10^{-14} m²/s [5.1]. Based on a diffusion coefficient of $D = 10^{-13}$ m²/s, depths of penetration into concretes containing 200 and 500 kg cement per m³ of 3.8 and 0.6 mm after 3 years' exposure to solutions with SO_4 concentrations of 0.1 and 1% respectively were calculated [5.1].

5.4 Alkali–aggregate reactions (AAR)

Special types of siliceous and in rare cases also dolomite aggregate may not behave as an inert material in concrete but enter chemical reactions with alkalis, e.g. sodium and potassium. These alkalis either originate from the cement or additives used in manufacturing the concrete or from the environment: for example seawater or deicing salt.

5.4.1 ALKALI–SILICA REACTION (ASR)

Amorphous or poorly crystallized silica present in natural aggregates such as chert, flint, chalcedony and opaline sandstone may be attacked by high alkali concentrations, forming a gel-like compound. Disruptive forces then develop, causing excessive cracking of the concrete. Two theories have been presented to explain the expansive phase formation [5.55]:

1. The gel itself causes an increase in solid volume, and pressure develops as the gel fills the pore spaces of the matrix.
2. Osmotic pressures develop as the silica gel absorbs water from the surrounding matrix.

Although ASR can only develop on concrete mixes containing reactive aggregates and at the same time sufficiently high concentrations of alkalis, permeability of the concrete exerts an important effect on the development of degradation, because excessive gel formation requires a high mobility of alkali ions as well as of water. Furthermore, the reaction may be intensified upon ingress of alkali ions and water from the environment into the concrete.

No quantitative data are available on the relationship between transport characteristics of concrete and its resistance against ASR. However, extensive studies on the beneficial effect of mineral additions such as fly ash or silica fume as well as partial replacement of clinker with slag indicate that a dense and impermeable matrix may improve the resistance against ASR considerably [5.56]. It must be considered, though, that the hydration products of pozzolanic additions such as fly ash or silica fume or slag of PBFS cements may combine free alkalis either in chemical reactions or by adsorption at their large specific surface area. Thus, a significant contribution to the observed improvement may also originate from a lower concentration of free alkalis in these systems. Differences in the chemical composition therefore also account for the higher resistance, aside from a reduced perviousness [5.52, 5.53].

5.4.2 ALKALI–CARBONATE REACTION (ACR)

Dolomitic limestone may undergo de-dolomitization in the presence of strong alkaline solutions, thereby forming a gel-like $Mg(OH)_2$. According to Hudec [5.55] this reaction may expose finely dispersed dry clay particles formerly enclosed in the limestone. Upon access of moisture, the clay particles absorb water and their swelling is thought to cause disruptive forces.

In principle, for ACR similar considerations for the damage development should be valid as for ASR with respect to ion mobility and water permeability; however, quantitative data are not available. It is interesting to note, though, that the susceptibility of both amorphous siliceous aggregates to ASR and dolomite aggregates to ACR may be correlated to the permeability and water absorption of the aggregates themselves [5.55].

5.5 Frost resistance

Freezing of water in the pore system of concrete may cause pits in and delamination of near-surface regions of concrete sections if the free pore space is insufficient to accommodate the volume increase of ice formation. Thus, damage occurring on frost action mainly depends on the degree of saturation of the concrete pore system with water. Especially the coarse capillary pores of the matrix are of particular interest because water kept in these pores may already freeze at moderate frost action, and these pores may reach a critical degree of water saturation by capillary action, i.e. absorption of water [5.57, 5.58].

Hilsdorf *et al.* [5.29] presented data on relations between transport coefficients and frost resistance of concrete. A very close relation as given in Fig. 5.12 was found between the absorption rate for water and the weight loss of concrete cubes after 60 freeze/thawing cycles. The different concrete series tested were manufactured with either ordinary Portland cement or Portland blastfurnace slag cement with a slag cement of 65% by mass. The concretes were non-air-entrained.

In these investigations the specimens for the freeze/thaw tests and the absorption measurements were subjected to the same preconditioning regime for 56 days. Therefore, the absorption tests could monitor the uptake of water for the different concrete test series during freezing and thawing.

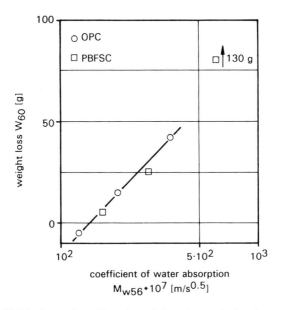

Fig. 5.12. Weight loss after 60 cycles of freezing and thawing versus absorption rate of concrete [5.29].

The good correlation between the absorption data and the observed frost resistance can be attributed to the fact that the rate of absorption describes the time required for a given concrete to reach a critical degree of saturation. If a critical degree of saturation is achieved the ice formation upon freezing will then induce frost damage.

In correlating the weight loss observed with the air permeability prior to freeze/thaw testing the observed tendency shows a decreasing frost resistance with increasing air permeability. However, for a given air permeability, concretes manufactured with PBFSC exhibit higher weight losses than OPC concretes (see Fig. 5.13). This behaviour was attributed to changes in the pore structure due to carbonation that could occur during preconditioning. The air permeability of the concrete specimens was tested at an age of 56 days. At the same age the freeze/thawing cycles were started.

Results on the oxygen permeability and frost resistance of different concretes presented by Gräf and Bonzel [5.37] are shown in Fig. 5.14. In this diagram the weight loss was recorded after 100 cycles of freezing and thawing.

The letters attributed to the symbols in this diagram describe the curing regime of the test specimens as given earlier; compare for section 5.2.1 and Fig. 5.5.

No data are available on correlations between frost/deicing salt resistance and transport coefficients. For air-entrained concretes interruption of the capillary pore system should be reflected in the absorption rate for

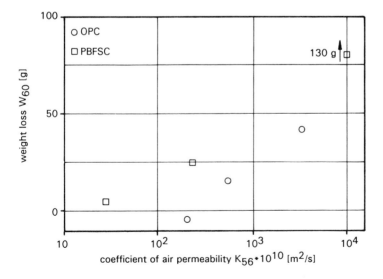

Fig. 5.13. Weight loss after 60 cycles of freezing and thawing versus air permeability of concrete [5.29].

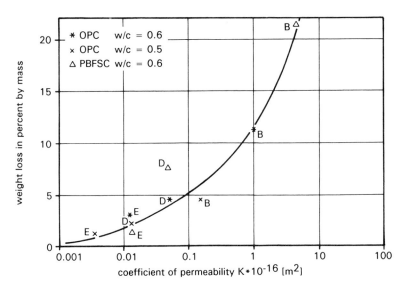

Fig. 5.14. Weight loss after 100 cycles of freezing and thawing versus air permeability [5.37].

water; however, these air voids may not exert a significant influence on the gas permeability (cf. Fig. 5.9). Therefore, gas permeability may not be a suitable indication of the frost and deicing salt resistance of air-entrained concretes.

In a testing concept developed by Fagerlund [5.59, 5.60] the critical degree of saturation serves as a criterion to evaluate the frost resistance of concrete. The critical degree of saturation is defined as the moisture content of concrete that causes a marked damage already after a few freeze/thawing cycles. If the critical degree of saturation is known, tests on the capillary absorption then will show if and after which duration of wetting the critical degree of saturation can be acquired.

In experimental investigations [5.60] on capillary absorption, Fagerlund used concrete discs, diameter 95–100 mm, height 23–26 mm, which were preconditioned by oven drying at 47°C for 12 days. The absorption tests showed a bilinear relation between the take-up of water and the square root of time. After the rising moisture front had reached the top of the concrete disc a further take-up of water occurred, at a much lower rate, however. The moisture content associated with a critical degree of saturation was attained within this second absorption phase.

Although related to transport characteristics this concept is not based on a time rate of flow but on a total absorption gained after a given period. The absorption rate is an indicator, however, when a critical stage can be reached.

Therefore, the result of this evaluation does not give information on the frost resistance of concrete in general but must be related to a specific exposure condition by comparing the duration of wetting periods in service with the time required to reach a critical degree of saturation in absorption tests.

5.6 Leaching

Leaching of constituents of hcp is a corrosion mechanism that involves the transport of species out of the matrix. For this mechanism a high water content of the matrix is required in order to support the dissolution of soluble compounds and enable the diffusion of ions in the liquid phase. The diffusion rate is then controlled by the concentration gradient that is effective across a given concrete section as well as the diffusion coefficient for individual ions describing their mobility.

A high mobility has been attributed especially to sodium, potassium as well as hydroxyl ions. An excessive dissolution and removal of calcium hydroxide may weaken the structure of hcp and initiate the decomposition of other hydration products [5.52].

Very porous and thus very pervious concrete may undergo severe leaching attack upon sustained permeation of water across a concrete section. Rather than diffusion the removal of hcp compounds then represents a convective mechanism with the capillary flow of water acting as a vehicle. Structural concrete of normal density usually achieves a sufficiently low water permeability to resist such extensive permeation of water. However, a high water permeability is frequently observed at work joints, cracks, and inhomogeneities such as sections with honeycombed structure.

In recent years leaching of cementitious materials has gained further importance in the field of waste management. Aside from low active nuclear wastes, industrial wastes with toxic compounds are also frequently solidified with cement for isolation and disposal [5.52]. At this point, however, the potentials of cementitious materials for fixation of waste or hazardous materials will not be discussed in more detail. However, it is pointed out that in these concepts the permeability of concrete is considered as a major criterion for evaluating the safety aspects of the disposal.

5.7 Soft water attack

The matrix-forming hydration products of the cement are soluble in water, although their solubility is comparatively low, and therefore most natural waters do not attack the paste. Owing to its low concentration of dissolved ions soft water favours the dissolution of the hydration products, which then may be carried out of the matrix by diffusion, a high concentration gradient acting as driving force, or they are removed from the reaction front by a convective flow, either through the pore system of the paste or at the surface of a concrete member.

The corrosiveness of penetrating soft water is often enhanced by dissolved carbon dioxide. The free carbon dioxide then increases the solubility of calcium hydroxide, calcium hydrates and calcium carbonate by the formation of calcium bicarbonates [5.61].

The reaction mechanism demonstrates that a severe attack may occur only provided there is a continuous supply of soft water with or without dissolved carbonic acid. This supply may prevail at the surfaces of concrete members or in interior sections if, owing to a very high water permeability of the concrete, the soft water penetrates easily through the concrete.

Although the water permeability of concrete is an important parameter controlling soft water attack, the concentration of soluble compounds as well as the solubility of the paste constituents are additional parameters of equal importance. In the presence of free carbonic acid the type of aggregate must also be considered because calcareous aggregates are vulnerable to dissolution too.

5.8 Acid attack

The major constituents of the hydrated cement paste matrix and eventually also the aggregates, e.g. calcareous materials, are dissolved by mineral acids as well as by organic acids, the latter usually being less aggressive than strong mineral acids. Organic acids may originate from the soil that is in contact with the foundation, from fruit and crops stored in silos and bunkers, from oil etc. Mineral acids usually originate from industrial processes where concrete is employed for retaining structures. However, also for natural exposure conditions the attack of mineral acids may occur, such as natural water with an excessive amount of dissolved carbon dioxide forming carbonic acid, or, as a result of the air pollution, sulphuric acid in rain, which then may exhibit pH values as low as pH $\approx$ 4 [5.62].

In standards [5.63, 5.64] the aggressiveness of acids is judged according to their pH, and a very severe attack is usually expected for acids with pH < 4.5. It has been shown that additional parameters must be considered too [5.62, 5.65].

Except for the case of the weak carbonic acid permeating through very porous concrete as discussed in section 5.7 the attack of an acid is normally limited to the immediate surface layer of a concrete section where the dissolution of compounds soluble in the given acid takes place in the first instant. For the further progress of the reaction front, however, it is very important to distinguish whether or not the acid continuously exposes new concrete sections and a free supply of new acid is provided, or whether the reaction products or insoluble residues will remain in place, thus restricting the supply of new acid and eventually forming a protective layer.

Grube *et al.* [5.62, 5.65] have observed that under the attack of carbonic acid the CSH of the hydrated cement paste matrix is dissolved with an insoluble residue of hydrous silicates. These silicates may provide a certain protective effect for concrete layers underneath as long as the gel-like material remains in place. However, the protective material is very weak and may easily be wiped off or be removed by mechanical forces, or may be carried away by the flow of the aggressive liquid along the surface. Then, the protective effect is lost and a new concrete surface is exposed directly to the aggressive agent.

Thus, two principal cases may be distinguished:

1. No protective layer is formed and a free access of acid of constant concentration is provided: in this case the propagation of the reaction front corresponds to the recess of the concrete surface. The velocity of the process is controlled by the specific rate of dissolution of the concrete constituents. The progress of the reaction front with time is then represented by the straight line in Fig. 5.15.
2. The acid attack results in the formation of an insoluble and stable layer, which covers the surface of the concrete section; thus the aggressive ions of the acid must diffuse through this layer before they reach unreacted concrete constituents. With time the thickness of the surface layer increases and the corrosion reaction becomes diffusion controlled; i.e. the propagation of the reaction front is determined by the supply of acid to the reaction front. Basically, this process is the same as the development of the depth of carbonation with time and therefore, considering corresponding simplifications, the progress of the reaction front is described by the $\sqrt{t}$ function in Fig. 5.15.

If the protective layer is periodically removed the progress of the reaction front may be described by a series of $\sqrt{t}$ relations in consecutive time intervals.

It is obvious that in the first type of attack, where no protective layer is formed or such layer is immediately removed, no transport characteristics of the concrete itself will have any direct effect on the corrosion process, because the aggressive agent will act on the surface and is not required to penetrate into the concrete.

In the second case the acid has to penetrate through the protective surface layer. Therefore, the corrosion rate is controlled by the transport properties of the corrosion products rather than by the transport properties of the original concrete. Although this is similar to the process of carbonation it is doubtful in the case of acid attack whether the transport

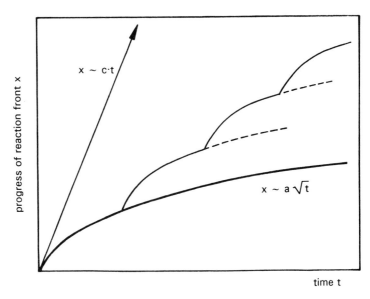

Fig. 5.15. Propagation of the reaction front with time for acid attack with and without the formation of a protective surface layer [5.65].

properties of the weak gel-like material correlate to the original concrete in any meaningful way.

5.9 Abrasion resistance

The surface zones of concrete members may undergo wear and abrasion. This corrosion mechanism is of general importance for all concrete pavements for foot and vehicle traffic such as roads and industrial floors, but structural members, e.g. the walls of silos and bins, may also suffer from abrasion. In hydraulic systems concrete surfaces may be worn off by the flow of water carrying sand particles and gravel. In such hydraulic systems a special case of damage referred to as cavitation erosion is caused by a high velocity flow of water such that negative pressures occur.

Except for cavitation erosion the abrasion of concrete surfaces is caused by a sliding or scraping action or an indentation of hard particles on the concrete surface. In a theoretical analysis of this problem Kunterding [5.66] applied finite element methods to calculate the stresses that are induced in the near-surface regions of concrete by these actions. His results show that locally tensile and shear stresses develop, which finally lead to failure if the local strength is exceeded. Depending on the size of particle, the location of impact, and type of motion relative to the concrete surface, the strength of the cement paste matrix, the strength and hardness of fine or coarse aggregates, and paste–aggregate bond strength may be the parameter controlling the abrasion resistance of concrete.

An abrasive attack is often superimposed by physical and chemical attack, e.g. if hot cement clinker is loaded into concrete bins or organic acids from farm silage penetrates into the concrete.

Numerous test methods exist for the evaluation of the abrasion resistance of concrete [5.67–5.69]; their suitability, however, is mostly limited to a specific type of abrasive attack, and none of the test methods is non-destructive.

As discussed before, abrasion resistance depends on the strength characteristics of concrete constituents. Although the macroscopic strength of concrete, e.g. compressive strength measured on cubes or cylinders, may not provide satisfactory indications, methods such as microhardness testing on the matrix and/or aggregate phase of the concrete may give better indications for the abrasion resistance [5.66].

Similarly, transport coefficients of the concrete may correlate well with its abrasion resistance because they are a sensitive measure for porosity, pore structure and microdefects which also control the abrasion. However, they may not be a characteristic for those abrasive attacks that are controlled by the hardness of the coarse aggregates.

In an experimental investigation on the effect of curing on the abrasion resistance of concrete, Senbetta [5.70] found that the absorption of water closely correlates with the quality of curing and accordingly also monitors the effect of curing on the abrasion resistance of concrete. In his experiments the abrasion resistance of concrete was evaluated by the rotating disc abrasion test according to ASTM C779 [5.69] as well as sand blasting test according to ASTM C418 [5.68]. The absorption was measured on mortar companion specimens with the same curing history as the concrete specimens. In Fig. 5.16 the abrasion of the different test series is given for the two test methods as a function of the absorptivity.

In a comprehensive study on the properties of the near-surface regions of a concrete section, Dhir [5.41] also investigated the abrasion resistance of different concrete mixes with different curing regimes. The abrasion resistance was assessed by a rolling wheel type machine [5.41] and for the initial surface absorption the take-up of water after 10 min was used. The concrete slabs for the absorption measurements had been predried at 105°C prior to the absorption measurement in order to improve the repeatability of the results. Details of the absorption tests are given in [5.38, 5.39]. In Fig. 5.17 the observed correlation between the depth of abrasion and the absorption behaviour of different concrete test series is shown. The concrete series tested differed in the mix composition, maximum size of aggregates, type and concentration of admixtures and curing. Details of the test programme are given in [5.38, 5.39].

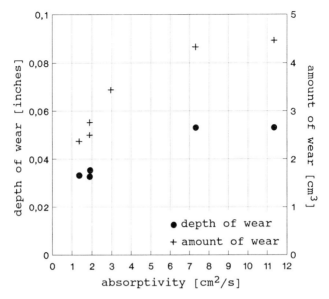

Fig. 5.16. Depth of wear according to ASTM C779 and amount of wear according to ASTM C418 of mortar specimens with different absorptivity [5.70].

Based on these results, Dhir considers the ISA test as a reliable and effective tool for the assessment of abrasion resistance. For on-site measurements, air drying of the concrete surface prior to the ISA test is recommended [5.41].

The experimental programme of Dhir also included measurements on the air permeability of the different concretes. As test specimens, drilled cores were taken from the slabs that were manufactured for the abrasion tests. The air permeability was measured on concrete discs, diameter 100 or 50 mm, thickness 50 mm [5.41]. Prior to the measurement the samples were oven-dried at 105°C.

In Fig. 5.18 the observed correlation between the abrasion depth and the coefficient of air permeability is given. Both axes of the diagram are given in a logarithmic scale. Although there is a clear tendency, the correlation is less pronounced than for the ISA test.

Further to capillary suction and air permeability the diffusion coefficient for water vapour was also considered. Similar to the results obtained on the air permeability, a clear tendency was observed; however, the correlation was weaker than for the ISA test. Figure 5.19 gives the observed correlation between the depth of abrasion and the diffusion coefficient for water vapour. Both axes of the diagram have a logarithmic scale.

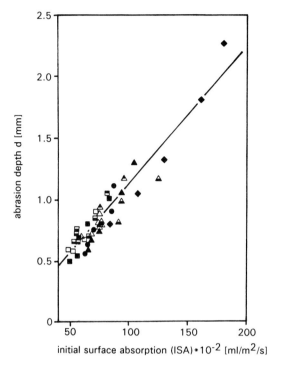

Fig. 5.17. Relationship between initial surface absorption and abrasion resistance [5.41].

5.10 Chloride ingress

The ingress of chlorides into concrete may be of multiple importance for the durability of concrete and reinforced concrete structures, and the most important implications refer to the combined action with frost as well as corrosion of the reinforcement. An overview of these mechanisms is given in Chapter 6.

The service life of a reinforced concrete structure may be defined as the time up to the onset of reinforcement corrosion. Similar to the progress of carbonation the ingress of chlorides, therefore, is of particular interest for the assessment of the lifetime of a structure, and the depth of chloride penetration for a given exposure may be regarded as a durability characteristic of the concrete.

The perviousness of concrete for chlorides may be evaluated by the direct measurement of transport parameters such as diffusion coefficient, by monitoring chloride profiles moving into a concrete section, or by indirect methods such as the AASHTO RCPT. Details of these methods are given in section 9.6.

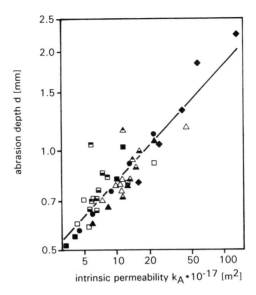

Fig. 5.18. Relationship between air permeability and abrasion resistance [5.41].

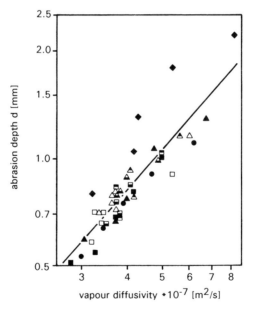

Fig. 5.19. Relationship between vapour diffusivity and abrasion resistance [5.41].

Additionally, limited information is available on the correlation between transport parameters for water or gases and the depth of chloride penetration.

As described earlier, reinforcement that is embedded in concrete of high alkalinity will undergo chloride-induced corrosion only after the chloride concentration in the adjacent concrete exceeds a certain threshold concentration. For the assessment of the corrosion protection of the reinforcement either concentration profiles for chlorides or the rate of progress of a given chloride concentration into the concrete section are of interest rather than the maximum depth of chloride ion penetration.

In [5.29] the penetration of chloride ions into concrete specimens was correlated with the air permeability of the concretes.

For the chloride penetration tests three different concrete mixes with two different Portland cements, $w/c = 0.45$ and 0.6 respectively, and air entrainment were manufactured and cylinders, diameter 100 mm, height 150 mm were cast. Curing under wet burlap was done for 1, 3 and 7 days. Subsequently, the concrete cylinders were sealed on the circumference to assure axial drying during subsequent storage in 20°C, 65% RH up to an age of 56 days. For the chloride penetration test the top face end of the cylinders was subjected to a 3% sodium chloride solution for a total of 56 days, exchanging the chloride solution every 2 to 3 weeks, however. The chloride profiles were taken by analysing drill powder, which was taken in a depth of 0–10 mm, 10–20 mm, 20–30 mm and 30–50 mm from the surface.

Specimens for the determination of the air permeability were subjected to the same curing and preconditioning regime up to the time of testing at an age of 56 days.

Figure 5.20 shows the depth x (mm) at which chloride concentrations of 0.4% and 1.0% by mass of cement had been observed on concretes with different coefficients of air permeability. The observed tendency indicates that for higher air permeabilities higher rates of chloride penetration can also be expected.

In [5.71] the chloride penetration into concrete is correlated to the take-up of water, which was observed for three different test methods:

1. capillary suction of water from the bottom face of a specimen, one-dimensional flow;
2. capillary suction from the top face of a specimen, three-dimensional flow;
3. depth of penetration of water under a pressure of 0.8 N/mm² for 24 h.

The penetration of chlorides was investigated in an immersion test, in which concrete specimens were placed into a 3% NaCl solution for 7 days and subsequently were kept in an indoor climate for 14 days for drying.

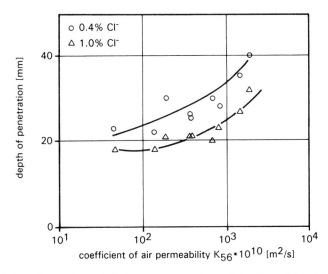

Fig. 5.20. Depth at which chloride concentrations of 0.4% and 1.0% by mass of the cement were observed on concrete specimens with different air permeabilities [5.29].

A total of 17 immersion/drying cycles were performed. Figure 5.21 shows the chloride ion concentration observed in a depth of 20 mm from the surface in relation to the take-up of water (1, 2) and depth of penetration of water (3), respectively.

These results demonstrate that a higher absorption rate as well as a higher depth of penetration of water in general is associated with a faster ingress of chlorides. However, for a given transport characteristic the concrete composition also plays an important role, as can be deduced from the differences between series containing fly ash or silica.

5.11 Corrosion of reinforcement

In the alkaline environment of concrete the reinforcement is protected against anodic iron dissolution by a passive layer on the steel surface. The drop in pH of the concrete pore water due to carbonation as well as chloride concentrations exceeding the threshold value will cause an extended or local breakdown of the passive layer, thus encouraging the anodic step of iron dissolution. Corrosion of the reinforcement, however, furthermore requires a sufficient supply of oxygen to support the cathodic corrosion reaction as well as moisture to serve as an electrolyte of low resistance. These two additional requirements may become counteracting in extreme dry or wet exposure conditions. For concrete continuously saturated with

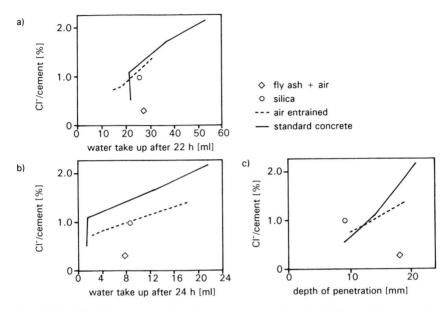

Fig. 5.21. Chloride concentration in concrete at a depth of 20 mm for concrete series with different transport characteristics [5.71].

water a low ohmic resistance between anodic and cathodic sites is given; however, access of oxygen may be limited to support the cathodic process. On the other hand, the open pore system of dry concrete is permeable for oxygen but in this stage lack of electrolyte will not allow for extensive corrosion.

Based on these mechanisms the ohmic resistance as well as the oxygen permeability of the concrete could yield a quantitative indication of potential corrosion rates. Further information on the resistivity of concrete can be found in [5.72].

Andrade *et al.* [5.73] investigated the availability of oxygen at the cathodic area of reinforcement in relation to the corrosion current, using as driving force for the flow of oxygen electrochemical polarization of electrodes embedded in cement paste, mortar and concretes. These materials had been preconditioned in 50% RH, 100% RH, or immersed in water in order to attain different moisture concentration.

It was found that for specimens conditioned at 50% and 100% RH respectively, the oxygen availability at the steel surface exceeded the amount needed for the corrosion observed. Only for specimens totally immersed in water was the supply of oxygen in accordance with the corrosion rate. In this state the corrosion is controlled by the cathodic reaction, and estimates of corrosion rates may be given on the basis of oxygen permeability.

5.12 Conclusions

In the present chapter an attempt was made to collect available information on the correlation between various transport mechanisms of concrete and its resistance against corrosive attacks. Although a close correlation had been postulated early already an urgent need for corresponding research in order to reveal quantitative data had been identified only in the 1980s [5.6]. Since then, some progress has been made resulting in correlations between most common corrosive attacks and transport characteristics.

The approach of assessing concrete durability by means of its transport characteristics is based on the assumption that corrosive mechanisms involve the movement of substances such as aggressive ions, gases or just water into the concrete, through concrete, or, as far as leaching is concerned, also out of the concrete. This assumption holds true for the majority of corrosive attacks encountered for concrete elements in service in various exposure conditions. Nevertheless, transport characteristics describe only one part of a corrosion mechanism, e.g. the supply of hazardous material at a reaction front. However, transport characteristics do not provide information on the rate and extent of a reaction or the total amount of substance transformed. Although an important parameter, concrete transport characteristics may therefore not be an overall criterion for concrete durability sufficient to describe the resistance of concrete against various chemical and physical attacks. Simultaneously, parameters representing the concrete composition or the composition of the cement may be of equal importance. Also, physical characteristics of the concrete such as its pore structure and pore volume must be considered.

In the reviewed literature quantitative or qualitative data on correlations between transport parameters of concrete and the following durability characteristics were presented:

Carbonation

Theoretically, the process of carbonation is controlled by the diffusion of carbon dioxide through the already carbonated surface zone towards the reaction front. The carbonation rate of concrete has been closely correlated with the gas permeability of concrete as well as with the capillary absorption of water for a particular concrete. Additional information on the type of cement used as well as type and amount of mineral additions should be available. The most important parameter for the carbonation rate of concrete exposed to outdoor conditions, however, is given by the moisture concentration of the concrete in service: pores saturated with water are impermeable for the gaseous CO_2, and therefore the diffusion coefficient for carbon dioxide is a function of the water content of the concrete.

Sulphate attack

The sulphate attack depends on two essential parameters, i.e. a sufficient supply of sulphate ions to the reaction sites and simultaneously a sufficient amount of vulnerable material available in the concrete to enter reactions with the sulphate ions. The transport of ions has been investigated by means of capillary absorption tests with sulphate solutions, and close correlations with the resistance of the concretes were observed. However, from the corrosion mechanism it follows that additional information on the type of the cement used must be available. Also, the effect of mineral additions deserves special attention because not only the transport characteristics will be altered but the chemical behaviour of the hydration products is changed simultaneously.

Alkali–aggregate reaction

In principle, the reaction partners for alkali–aggregate reactions are present in the concrete itself. Nevertheless, the perviousness of concrete is considered as an important parameter because a transport of water and ions to the reaction sites as well as a transport of the gel material formed, i.e. alkali–silica gel or magnesium hydroxide, away from the reaction sites must occur.

In those cases where the alkali–aggregate reactions are supported by an external source of alkalis, e.g. seawater, as well as increasing the moisture content of the concrete by the ingress of water, the importance of transport parameters becomes evident. As additional factors the vulnerability and the reactivity of the aggregates as well as the concentration of alkalis in the cement must be considered.

Although the correlation between the perviousness of concrete and its resistance against alkali–aggregate attack is generally accepted, no quantitative data are available on these correlations.

Frost resistance

Damage of concrete due to frost action will develop if the pore system of concrete is saturated with water to a critical degree. This high water content may be achieved only after take-up of water from the environment, e.g. by capillary suction.

Experimental results showed that the frost resistance of non-air-entrained concretes correlated closely with the capillary absorption rate, but also with the coefficient of air permeability.

No data are available on the correlation between transport characteristics and the frost and deicing salt resistance of air-entrained concretes. Although in experiments the presence of a protective air void system was reflected in a reduction of the capillary suction of the concrete, characteristics of the air void system such as volume of air voids, spacing factor etc. must be known to evaluate the potential durability of a given concrete.

Leaching

The corrosion mechanism involves the transport of species out of concrete, either by diffusion of ions or, upon penetration of water, also by a convective flow. Thus the close correlation between the resistance against leaching and transport characteristics such as water permeability or ion diffusivity is evident.

No quantitative data are available and additional effects should be expected from the mobility of the leached species.

Soft water

The attack of soft water follows the same principles as leaching attack and therefore the same correlations exist.

Acid attack

The attack of acids may lead to two distinct corrosion processes:

- If no protective surface layer forms during the corrosion process the progress of the reaction front is solely controlled by the dissolution behaviour of the material directly exposed to the acid.
- If a corrosion product remains in place, thus forming a surface coating, the progress of corrosion turns to be controlled by the supply of acid at the reaction front. Thus, the diffusivity of the corrosion product will be a characteristic value.

In the case of acid attack no correlation between the durability of concrete and its transport characteristics exist which is supported by the physico-chemical mechanism of attack.

Abrasion

Abrasion of concrete is caused by mechanical stresses that locally exceed the tensile or shear strength of concrete. Depending on the type of attack the damage may be limited in the hydrated cement paste matrix or also the coarse aggregates may be worn off. However, in no case is a transport process involved in abrasion.

Nevertheless, experiments have proved that those abrasive attacks that are focused on the hydrated cement paste matrix may closely correlate with a variety of transport characteristics, such as gas permeability, capillary suction of water and even diffusion of water vapour. These empirical correlations may be explained with the help of the porosity and pore structure of the hydrated cement paste matrix, which controls in a similar manner both strength and transport characteristics.

Chloride ingress

The ingress of chlorides into concrete is either due to a diffusion process of chloride ions or is caused by the capillary suction of a chloride solution. These mechanisms become active for very different moisture states of the concrete.

In experimental investigations it has been shown that the chloride ingress may be correlated to the gas permeability of concrete; a closer correlation, however, is observed with the capillary suction properties of concrete.

To a certain degree penetrating chloride ions are immobilized at the hydration products. The capacity for immobilization depends on the type of cement as well as on the type and amount of additions used. Therefore, as additional information data on the cement composition and materials used for concrete production are necessary.

Reinforcement corrosion

With respect to transport phenomena the corrosion of reinforcement should depend on the supply of oxygen at the cathodic sites and the availability of water to act as an electrolyte.

Whereas for normal outdoor exposure of concrete the water content in equilibrium with the atmosphere is usually sufficient to act as an electrolyte only for concretes submerged in water was the reinforcement corrosion controlled by the oxygen supply. No direct correlations between the corrosion rate of the reinforcement and transport parameters of concrete have been established yet.

5.13 References

5.1. Oberholster, R.E. (1986) Pore structure, permeability and diffusivity of hardened cement paste and concrete in relation to durability: status and prospects, in *Proc. 8th Int. Congress Chemistry of Cement*, Rio de Janeiro.
5.2. Feldman, R.F. (1986) Pore structure, permeability and diffusivity as related to durability, in *Proc. 8th Int. Congress Chemistry of Cement*, Rio de Janeiro.
5.3. Hilsdorf, H.K. (1989) Durability of concrete – a measurable quantity? IABSE Symposium, Lisbon, *IABSE Report*, Vol. 57, No. 1, pp. 111 23.
5.4. Levitt, M. (1969) An assessment of the durability of concrete by ISAT, in *Proc. RILEM Symposium on Durability of Concrete, Prague.*
5.5. Figg, J.W. (1973) Methods of measuring the air and water permeability of concrete. *Magazine of Concrete Research*, Vol. 25, No. 85, December, pp. 213–219.
5.6. The Concrete Society (1985) *Permeability of concrete and its control*, Papers of a one-day conference, Tara Hotel, Kensington, London, 12 December.
5.7. Comité Euro-International du Beton – CEB (1991) CEB FIP Model Code 1990, Final Draft Bulletin d'Information No. 203, 204, 205, Lausanne.

5.8.   British Standards Institution. BS1881, Methods of testing concrete BS 368, Precast concrete flags, 1971. BS 390, Specification for precast concrete kerbs, channels, edgings and quadrants, 1979.

5.9.   DIN 1045 (1988) Beton und Stahlbeton – Bemessung und Ausführung.

5.10.   ENV 206 (1990) Concrete – Performance, production placing and compliance criteria.

5.11.   Comité Euro-International du Beton – CEB (1989) *Durable Concrete Structures – CEB Design Guide*, Bulletin d'Information No. 182, Lausanne.

5.12.   Comité Euro-International du Beton – CEB, Task group durability (1982) *Durability of concrete structures – state of the art report*, Bulletin d'Information No. 148, Paris.

5.13.   RILEM (1976) International Symposium on Carbonation of Concrete, Cement and Concrete Association, Wexham Springs, Slough, UK.

5.14.   Vesikari, E. (1989) Prediction of the service life of concrete structures, Finnish Contributions to RILEM 43rd General Council Meeting, Espoo.

5.15.   Meyer, A., Wierig, H.J. and Husman, K. (1967) Karbonatisierung von Schwerbeton, *Schriftenreihe des Deutschen Ausschusses für Stahlbeton*, Heft 1982, Berlin.

5.16.   Papadakis, V.G., Fardis, N.M. and Vayenas, C.G. (1992) Hydration and carbonation of pozzolanic cements. *ACI Materials Journal*, March/April, pp. 119–130.

5.17.   Papadakis, Y.G., Fardis, M.N. and Vayenas, C.G. (1992) Effect of composition, environmental factors and cement-lime mortar coating on concrete carbonation *Materials and Structures*, Vol. 25, pp. 293–304.

5.18.   Bakker, R.F.M. (1988) Initiation period, in *Corrosion of steel in concrete*, RILEM Report of the TC 60 CSC; Ed. Schießl, P., Chapman & Hall, pp. 25–55.

5.19.   Verbeck, G.J. *Carbonation of hydrated Portland cement*, ASTM Special Technical Publication No. 205.

5.20.   Kamimura, K. (1965) Changes in weight and dimensions in the drying and carbonation of Portland cement mortars. *Magazine of Concrete Research*, Vol. 17, No. 50.

5.21.   RILEM Committee CPC 18 (1985) Draft Recommendation, Measurement of hardened concrete carbonation depth. *Materials and Structures*, No. 102.

5.22.   Powers, T.C. (1962) A hypothesis on carbonation shrinkage. *Journal of the PCA Research and Development Laboratories*.

5.23.   Bier, Th.A. (1988) Karbonatisierung und Realkalisierung von Zementstein und Beton. Dissertation, Universität Karlsruhe.

5.24.   Kropp, J. (1983) Karbonatisierung und Transportvorgänge in Zementstein. Dissertation, Universität Karlsruhe.

5.25.   Schießl, P. (1976) Zur Frage der zulässigen Rißbreite und der erforderlichen Betondeckung im Stahlbetonbau unter besonderer Berücksichtigung der Karbonatisierungstiefe des Betons. *Schriftenreihe des Deutschen Ausschusses für Stahlbeton*, Heft 255, Berlin.

5.26.   Smolczyk, H.G. (1968) Written discussion (about carbonation evaluation with time on different cements), in *Proc. 5th Int. Symp. Chemistry of Cements*, Vol. III, Tokyo.

5.27.   Wierig, H.J. (1984) Longtime studies on the carbonation of concrete under normal outdoor exposure, RILEM Seminar *Durability of concrete structures under normal outdoor exposure*, Hannover, pp. 239–249.

5.28. Houst, Y. and Wittmann, F.H. (1986) The diffusion of carbon dioxide and oxygen in aerated concrete, in *2nd Int. Colloquium on Materials Science and Restoration*, 2–4 September, Esslingen, Germany.

5.29. Hilsdorf, H.K., Schönlin, K. and Burieke, F. (1992) *Dauerhaftigkeit von Betonen*, Institut für Massivbau und Baustofftechnologie, Universität Karlsruhe.

5.30. Paulmann, K. and Rostasy, F.S. (1990) *Praxisnahes Verfahren zur Beurteilung der Dichtigkeit oberflächennaher Betonschichten im Hinblick auf die Dauerhaftigkeit*, Institut für Baustoffe, Massivbau und Brandschutz, TU Braunschweig.

5.31. Bunte, D. and Rostasy, F.S. (1989) Test methods for on-site assessment of durability, IABSE Symposium, Lisbon.

5.32. Rostasy, F.S. and Bunte, D. (1989) Evaluation of on-site conditions and durability of concrete panels exposed to weather, IABSE Symposium, Lisbon.

5.33. Hardt, R. (1988) Einfluß einer Karbonatisierung auf die Permeabilität von Beton. Diplomarbeit am Institut für Massivbau und Baustofftechnologie, Universität Karlsruhe.

5.34. Osborne, G.J. (1989) Carbonation and permeability of blast furnace slag cement concretes from field structures, in *3rd International Conference Fly Ash, Silica Fume, Slag and Natural Pozzolans*, Trondheim, ACI SP-114.

5.35. Parrott, L. Water absorption in cover concrete. *Materials and Structures*, Vol. 25, No. 149, June, pp. 284–292.

5.36. Lawrence, C.D. (1989) Transport of oxygen through concrete. *Proceedings of the British Ceramics Society*, No. 35, pp. 227–293.

5.37. Gräf, H. and Bonzel, J. (1990) Über den Einfluß der Porosität des erhärteten Betons auf seine Gebrauchseigenschaften. *Beton*, Heft 7.

5.38. Dhir, R.K., Hewlett, P.C. and Chan, Y.N. (1987) Near-surface characteristics of concrete: assessment and development of in situ test methods. *Magazine of Concrete Research*, Vol. 39, No. 141.

5.39. Dhir, R.K., Hewlett, P.C. and Chan, Y.N. (1989) Near-surface characteristics of concrete: intrinsic permeability. *Magazine of Concrete Research*, Vol. 41, No. 147, June, pp. 87–97.

5.40. Dhir, R.K., Hewlett, P.C. and Chan, Y.N. (1989) Near-surface characteristics of concrete: prediction of carbonation resistance. *Magazine of Concrete Research*, Vol. 41, No. 148, September, pp. 122–128.

5.41. Dhir, R.K., Hewlett, P.C. and Chan, Y.N. (1991) Near-surface characteristics of concrete: abrasion resistance. *Materials and Structures*, Vol. 24.

5.42. Kikuchi, M. and Mukai, T. (1988) Carbonation of concrete containing fly ash coarse aggregate. *CAJ Review*.

5.43. CEB RILEM (1983) International Workshop, *Durability of Concrete Structures*, Workshop Report edited by S. Rostam, Copenhagen, May.

5.44. Goncalves, A. (1980) *Durability of High Strength Concrete*, Report to RILEM TC 116.

5.45 RILEM (1980) Tentative Recommendation. Recommended tests to measure the deterioration of stone and to assess the effectiveness of treatment methods, Part V, 'Durability Tests', Test No. V.1 'Crystallization test by total immersion'. *Materials and Structures*, Vol. 13, No. 75.

5.46. RILEM (1974) Tentative Recommendation No. 11.2, Absorption of water by capillarity. *Materials and Structures*, Vol. 7, No. 40.

5.47. RILEM (1979) Tentative Recommendation CPC 13.2, Test for permeability of porous concrete. *Materials and Structures*, Vol. 12, No. 69.

5.48. Van Aardt, J.H.P. (1955) *The Resistance of Concrete and Mortar to Chemical Attack*, National Building Research Institute, Pretoria, SA, Bulletin No. 13 and 17.

5.49. Tikalsky, P.J. and Carrasquillo, R.L. (1989) *The effect of fly ash on the sulfate resistance of concrete*, Center for Transportation Research, University of Texas at Austin, Research Report 481–5.

5.50. Kohno, K., Aihara, F. and Ohno, K. (1989) Relative durability properties and strengths of mortars containing finely ground silica and silica fume, in *3rd International Conference Fly Ash, Silica Fume, Slag and Natural Pozzolans*, Trondheim, ACI SP-114.

5.51. Yamato, T., Solda, M. and Emoto, Y. (1989) Chemical resistance of concrete containing condensed silica fume, in *3rd International Conference Fly Ash, Silica Fume, Slag and Natural Pozzolans*, Trondheim, ACI SP-114.

5.52. Roy, D.M. (1986) Mechanisms of cement paste degradation due to chemical and physical factors, in *Proc. 8th Int. Congress Chemistry of Cement*, Rio de Janeiro.

5.53. Uchikawa, H. (1986) Effect of blending components on hydration and structure formation, in *Proc. 8th International Congress Chemistry of Cement*, Rio de Janeiro.

5.54. Mehta, P.K. (1979) Pozzolanic and cementitious by-products in concrete – another look, in *3rd Int. Conference on Fly Ash, Silica Fume, Slag and Natural Pozzolans in Concrete*, Trondheim, V.M. Malhotra, (Ed.), ACI SP 114.

5.55. Hudec, P.P. (1990) Common factors affecting alkali reactivity and frost durability of aggregates, in *Durability of Building Materials and Components*, Proceedings Fifth International Conference Brighton, UK, J.M. Baker *et al.* (Eds), E & FN Spon, London.

5.56. Turriziani, R. (1986) Internal degradation of concrete: alkali aggregate reaction, reinforcement steel corrosion, in *Proc. 8th International Congress Chemistry of Cement*, Rio de Janeiro.

5.57. Setzer, M.J. (1977) Einfluß des Wassergehalts auf die Eigenschaften des erhärteten Betons. *Schriftenreihe des Deutschen Ausschusses für Stahlbeton*, Heft 280, Berlin.

5.58. Setzer, M.J. and Hartmann, V. (1991) Verbesserung der Frost Tausalzprüfung. *Betonwerk + Fertigteil Technik*, Heft 9.

5.59. Fagerlund, G. (1977) The critical degree of saturation method of assessing the freeze-thaw resistance of concrete. *Materials and Structures/Materiaux et Constructions*, Vol. 10, No. 58.

5.60. Fagerlund, G. (1977) The international cooperative test of the critical degree of saturation method of assessing the freeze/thaw resistance of concrete. *Materials and Structures/Materiaux et Constructions*, Vol. 10, No. 58.

5.61. Ballim, Y. and Alexander, M.G. (1990) Carbonic acid water attack of Portland cement based matrices, in *Protection of Concrete*, Proc. Int. Conference, University of Dundee, Scotland, 11 13 September, E & FN Spon, London, pp. 93–104.

5.62. Grube, H. and Rechenberg, W. (1987) Betonabtrag durch chemisch angreifende saure Wässer. *Beton*, Vol. 37, pp. 446–451 and 495–498.

5.63. DIN 4030 (1991) *Beurteilung betonabgreifender Wässer, Böden und Gase*, Beuth Verlag, Berlin.

5.64. ISO DP 9690 (1987) Classification of environmental exposure conditions for concrete and reinforced concrete structures.

5.65. Grube, H. and Rechenberg, W. (1989) Durability of concrete structures in acidic water. *Cement and Concrete Research*, Vol. 19, pp. 783–792.

5.66. Kunterding, R. (1991) Beanspruchung der Oberfläche von Stahlbetonsilos durch Schüttgüter. Dissertation, Universität Karlsruhe.

5.67. DIN 52108 (1988) Verschleißprüfung mit der Schleifscheibe nach Böhme.

5.68. ASTM-C-418-81. Standard Test Method of Abrasion Resistance of Concrete by Sand blasting.

5.69. ASTM-C-779-82. Standard Test Method of Abrasion Resistance of Horizontal Concrete Surfaces.

5.70. Senbetta, E. and Malchow, G. (1987) Studies on control of durability of concrete through proper curing, in *Concrete Durability - Katharine and Bryant Mather International Conference*, ACI SP-100, Detroit.

5.71. Johansson, L., Sundbom, S. and Woltze, K. (1989) *Permeabilitet, prouning och inverkan pa betongs beständighet*, CEB-report No. S-100 44, Cement och Betoninstituttet, Stockholm.

5.72. Schießl, P. Ed. (1988) *Corrosion of Steel in Concrete*, Report of RILEM TC 60-CSC, Chapman & Hall, London.

5.73. Andrade, C., Alonso, C., Rz-Maribona, I. and Garcia, M. (1989) Suitability of the measurement technique of oxygen permeability in order to predict corrosion rates of concrete rebars, Paul Klieger Conference, ACI, San Diego.

6 Chlorides in concrete

Jörg Kropp

Amongst the various exposure conditions that concrete and reinforced concrete structures may be subjected to during their service life, chlorides represent one of the most complex and eventually most hazardous attacks. Their impact on the performance of a structure is manifold. Correspondingly, much research work has been dedicated to revealing the corrosive effect of chlorides on structures.

It is not the intention of this chapter to present a comprehensive review of the literature on chloride attack, which is also beyond the scope of this state-of-the-art report. Instead, the following sections are limited to some fundamental considerations that are relevant to other topics of the report. In particular, this chapter should indicate the significance of chloride ingress for durability and provide additional information on the transport of chlorides and their measurement, thus supporting Chapters 2, 5 and 9, respectively.

6.1 Sources of chlorides

6.1.1 CHLORIDE-CONTAINING MATERIALS

Chlorides may already be introduced in the fresh concrete mix if the concrete-making materials are contaminated with chlorides. For use in reinforced or prestressed concrete structures the chloride concentrations in cements, mixing water, aggregates, and admixtures are therefore controlled, and maximum permissible concentrations are given in standards, approval documents and guidelines [6.1–6.3]. A survey on national and international documents stating permissible chloride concentrations in concrete-making materials is given in [6.4].

In most cases, however, excessive amounts of chloride in concrete originate from external sources. The penetration of chlorides into the concrete occurs by various transport mechanisms.

Performance Criteria for Concrete Durability, edited by J. Kropp and H.K. Hilsdorf.
Published in 1995 by E & FN Spon, London. ISBN 0 419 19880 6.

6.1.2 SEAWATER EXPOSURE

Marine structures are exposed to chlorides contained in the seawater. In view of the transport mechanisms involved in the penetration process distinction should be made between a sustained direct contact with seawater for submerged structures or elements, the cyclic exposure to seawater that takes place in the tidal zone, and elements subjected to splash and spray water only. Furthermore, coastal areas may have considerable chloride concentrations in the atmosphere, which may be deposited or washed out with rain then reaching the surface of structures [6.5, 6.6].

For submerged structures mainly ion diffusion, and to some extent also the permeation of the salt solutions, are responsible for the ingress of the chlorides into concrete. It must be considered, however, that concrete surface zones may form protective coatings with a low permeability due to ion exchange reactions with other compounds of seawater, resulting in films of $Mg(OH)_2$ and $CaCO_3$ [6.7, 6.8]. Therefore, the penetration rate of chlorides into these structures is often considerably lower than estimated from laboratory experiments, where no protective films can be formed due to the test method chosen.

For all concrete elements that may dry to some extent in the near-surface regions the ingress of chlorides into the concrete is supported by capillary absorption of the seawater upon direct contact. Capillary absorption thereby gains importance as the degree of drying between the individual wetting periods increases.

6.1.3 DEICING SALTS

Calcium chloride and sodium chloride are frequently used as deicing agents on concrete pavements for traffic areas such as roads, bridges, sidewalks, stairs and parking decks. Together with molten ice or snow a chloride solution is then formed, which eventually also reaches elements of a structure where no deicing salt has been applied directly. Carried by cars such salt solutions are brought into parking garage decks, and the substructure of bridges or structures adjacent to traffic areas is exposed to splashing or spray of salt solutions or the atmosphere [6.9, 6.10]. These salt solutions may also dry up, thereby increasing the salt concentration until finally leaving the crystalline salt on the concrete surface. Later on, a salt solution may be formed again if rainwater dissolves the salts in periods when no deicing agents are applied any more.

In all these cases the concrete surfaces are not continuously in contact with a salt solution. Alternating drying and wetting as well as exposure to salt solutions of varying concentration take place. Therefore, capillary absorption is the predominant mechanism that carries the chloride into the concrete. Repeated cycles of capillary absorption of salt solution and subsequent drying then increase the salt concentration in the concrete surface zone. Depending on the moisture concentration of the concrete the chlorides will penetrate into deeper sections by diffusion. Diffusion may also be responsible for the redistribution and flattening of steep chloride concentration profiles in near-surface regions of a concrete element [6.11–6.13].

6.1.4 FIRE

Aside from its immediate impact on the strength and serviceability of structural elements fire may additionally provoke a secondary corrosive effect on reinforced concrete structures, which results from the thermal decomposition of organic compounds. As one of the most important organic materials PVC starts to decompose at temperatures exceeding 80–90°C thereby liberating gaseous HCl. With increasing temperature the amount of HCl liberated increases, and decomposition is almost completed at temperatures around 300°C.

These temperature ranges and the amount of HCl liberated may vary depending on the type of PVC; nevertheless, temperatures occurring in the case of fire will be sufficient for decomposition of PVC. The gaseous HCl is carried away from the centre of the fire. Upon access of moisture the gas dissolves in the water to form hydrochloric acid of pH as low as 1 [6.14]. The acid then eventually condenses at colder sites on the surface of elements. In reactions with concrete compounds, calcium chloride ($CaCl_2$) is formed, which dissolves in water. The chloride ions then penetrate into deeper sections of the concrete, either by capillary absorption or by diffusion.

PVC is a widespread material with a high chlorine content. It must be considered, however, that other organic resins may contain chlorine additions too, in general for controlling the inflammability of the material. Chloride contamination of concrete elements therefore often occurs as a consequence of fire [6.14, 6.15].

6.1.5 OTHER SOURCES

Chlorides may also be present in groundwater or the soil so that foundations or buried structures or elements can be subject to chloride ions. Also, industrial processes may give rise to chloride attack. In a humid environment reinforced concrete elements should not be in contact with other building materials that may eventually liberate soluble chlorides [6.4].

6.2   **Significance of chlorides for the durability of structures**

With regard to the durability of concrete and reinforced concrete structures two distinctly different mechanisms for the corrosive action of chlorides must be distinguished. In combination with frost, chlorides contained in deicing salts cause a mainly physical attack on the concrete itself, which is much more severe than frost alone. The second corrosive action of chlorides is related to the reinforcement, which undergoes serious corrosion if sufficient oxygen and moisture are available.

6.2.1 FROST AND DEICING SALT ATTACK

The resistance of concrete against the action of frost is primarily controlled by its degree of moisture saturation. Water increases its volume by approximately 9% upon ice formation. Therefore, disruptive forces develop if the expansion of ice formation is not accommodated by empty pores. Also, the rate of cooling is important because expanding ice exerts a pressure on the remaining liquid water, which will then migrate to unfrozen sites. For high freezing rates, however, this transport of water may not be fast enough to prevent disruption. Nevertheless, concretes with a high resistance to freezing and thawing can be produced for exposure conditions where the concrete cannot reach moisture concentrations close to saturation [6.16–6.19].

The damage mechanisms for frost in combination with deicing agents are much more complex, and no unique theory is generally accepted yet. Additional effects due to the application of the deicing salt are reported, such as steep temperature gradients with high cooling rates occurring right after the salt application, a layered freezing and thawing of concrete sections, an increase of the water content in the pore system due to osmotic effects, crystallization pressure of the salt as well as a chemical attack of high salt concentrations.

These mechanisms demonstrate that the simultaneous attack of frost and deicing salt is much more complex and much more severe than frost alone. Characteristics of the pore structure and transport parameters play a decisive role in controlling the frost resistance. Air entrainment of the matrix has to provide pores which act as expansion space; therefore, these pores are not allowed to fill with water under the given exposure. In general a dense concrete is required with a low take-up of water as well as a low penetration rate for chlorides. There are indications, however, that a protective air entrainment with normal air content is ineffective for concretes with very low permeability, because in these systems the distances between the individual air voids are too large to allow for a rapid pressure release during freezing at the given low permeability [6.20].

On the other hand, very dense concretes, manufactured for example with silica fume additions, may never reach a critical degree of saturation. These concretes then exhibit a high resistance against freezing and thawing even without air entrainment [6.21].

6.2.2 CORROSION OF REINFORCEMENT

The steel reinforcement embedded in concrete is protected against corrosion by the passivation of the steel surface due to the high alkalinity of the concrete pore water. While carbonation encourages corrosion in the course of neutralization of the hydration products so that the passive layer becomes unstable, free chloride ions dissolved in the pore water will locally destroy the passive film, thus causing anodic iron dissolution even in a highly alkaline environment [6.22].

Chloride-induced corrosion of the reinforcing bars may form general corrosion if high chloride levels are spread over large areas of the steel surface; however, often pitting corrosion is observed. This type of corrosion occurs due to a local breakdown of the passive layer forming a very small anode. Owing to a sufficient supply of oxygen, large areas of the reinforcement surface remain cathodic. This particular corrosion element causes rapid iron dissolution at the anode, leading to deep pits and correspondingly a considerable reduction of the load-bearing cross-section of the bar.

In the electrochemical corrosion reactions, the chloride ions are not consumed, but may form only intermediate corrosion products, such as $FeCl_2$. Subsequent hydrolysis liberates the chloride ion again, and at the same time hydrogen ions are set free. This reduces the pH at anodic sites, and at cathodic sites the pH may raise because of increasing $(OH)^-$ formation. Hence the conditions become more favourable for corrosion, and high corrosion rates are generated. Thus pitting corrosion is self-supporting.

The individual components and reactions in the corrosion element are shown in Fig. 6.1, which is taken from the report of RILEM TC 60-CSC, *Corrosion of Steel in Concrete* [6.22]. This gives a comprehensive survey of chloride-induced corrosion of reinforcement.

Pitting corrosion is of particular importance for brittle materials such as prestressing tendons or wires, where the pits may act as notches. It must be considered also that some types of so-called stainless steel are vulnerable to chloride-induced pitting corrosion [6.23]. Metal coatings for corrosion protection, such as zinc, will undergo similar anodic corrosion [6.24].

6.2.3 CHEMICAL ATTACK

Salt solutions of high concentrations are also reported to cause a chemical attack on the hydrated cement paste. Upon access of sodium chloride

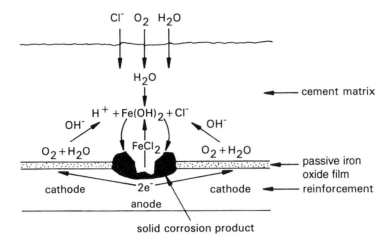

Fig. 6.1. Corrosion cell for chloride-induced pitting corrosion [6.22].

solutions the well-known Friedel's salt is formed: however, no deleterious effect is attributed to this reaction product. In contrast, solutions with high concentrations of calcium chloride or magnesium chloride may cause a chemical attack accompanied by a drop of the pore water pH and a disruption of the matrix. The damage mechanism is explained with the formation of expansive oxychlorides from portlandite [6.48].

6.3 Types of chloride in concrete

With respect to the individual corrosion phenomena as well as for the transport processes two types of chloride in concrete must be distinguished. These are the free chloride ions dissolved in the concrete pore solution, and chloride ions that are combined in or bound to different hydration products of the cement. Both types of chloride normally exist simultaneously to maintain chemical equilibrium [6.25, 6.26]. For transport processes as well as for corrosive actions on the reinforcement, however, only the free chloride ions in the pore solution are of importance, and knowledge of the total chloride content of concrete is often insufficient unless the relation between the free chloride ion concentration and the combined chloride concentration is known. The corresponding ratio is usually refered to as the binding capacity [6.12].

It is well established that the binding capacity for chlorides depends on the chemical composition of the cement, especially the content of C_3A in the clinker and the type and concentration of blending compounds such as ggbfs or fly ash, because they will affect both the absorption of the chloride ions by chemical reactions as well as the adsorption at the internal surface area. In this respect the effect of slag or pozzolana is discussed controversially. Blending Portland cement clinker with these materials lowers the concentration of C_3A, which is the most important compound for chemical binding of the chloride ion to form Friedel's salt, $3CaO.Al_2O_3$. $CaCl_2.10H_2O$.

In contrast, these materials increase the formation of CSH and CSH-like phases, which bind chlorides by adsorption due to surface forces. In the literature, results can be found evidencing a lower binding capacity for blended cements. However, other results suggest an increased binding, especially for cements with high slag concentrations [6.27 6.29].

In Fig. 6.2 the bound and therefore immobilized chloride concentration is plotted against the total chloride concentration of concretes that were manufactured with different types of cement. For low chloride concentrations the majority of chlorides are combined but the capability for binding will reach a limit for higher chloride concentrations. According to Fig. 6.2, this limit also depends on the type and composition of the cement used.

The different lines in Fig. 6.2 may be regarded as **sorption isotherms** for chlorides in concrete, showing two distinctly different ranges. In the first range, the binding of chlorides is proportional to the total chloride content as represented by the slope of the straight lines. Upon increasing chloride ingress the binding capacity is exhausted and no further increase in combined chlorides is possible. This state is indicated by the horizontal branches of the straight lines. In principle this behaviour corresponds to

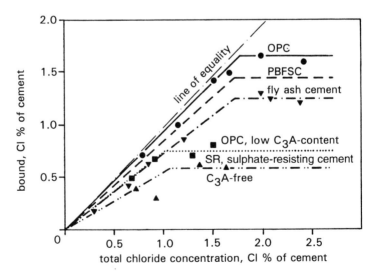

Fig. 6.2. Relation between bound chloride content and total chloride content of different concretes according to [6.32].

a Langmuir type of sorption isotherm, e.g. describing a monolayer adsorption. In the case of chlorides in concrete, however, binding by adsorption is additionally supported by chemisorption. Nevertheless, a Langmuir type of sorption isotherm may be used in prediction models for chloride ingress considering the immobilization of the chlorides (see ref. 147 in [6.28]).

Tuutti [6.12] showed that a linear relationship exists between the free chloride ion concentration in the pore water and the combined chloride concentration in hardened cement paste (hcp), and for the ratio of free chloride to combined chloride a range of 0.7 to 3.8 has been reported.

For the results given in Fig. 6.2 as well as in Tuutti's experiments, chloride was added to the fresh concrete mixes, and absorption of the chlorides could therefore take place during the formation of the hydration products. It has been reported, however, that the composition of the pore solution may differ depending on whether chlorides had been added to the mix or had penetrated into the hardened concrete [6.30], the latter case being of practical importance for concrete structures in service.

Further effects result from the type of chloride as well as from the moisture concentration of the concrete. According to Tuutti [6.12], chlorides originating from $CaCl_2$ have been combined to a higher extent than those resulting from KCl. Furthermore, evaporation of the pore water upon drying will increase the free chloride concentration in the remaining pore water.

It is important to note that carbonation of the hydration products causes an extensive decomposition of the hydration products, and also the chloride-bearing hydration products will be destroyed. Friedel's salt decomposes into calcium carbonate and aluminum oxide, thus liberating chloride and water. Also, the CSH phases are decomposed, thereby losing their capacity to combine chlorides. In a carbonated matrix, therefore, the total chloride content is likely to prevail as free chloride ions. This implies that carbonation sets free the formerly immobilized chloride ions, which then contribute to the migration. Additionally, the deleterious effect of the increased concentration of free chloride ions on the steel corrosion must be considered [6.22].

6.4 Critical chloride concentration

Chlorides are widespread and can be detected in small concentrations practically everywhere: in materials, natural exposure conditions and industrial processes as well. Neither for the concrete itself nor for the reinforcement is a chloride-free condition realistic. Low concentrations of chlorides must be taken into account already due to permissible impurities of the raw materials for concrete production, and in general these low chloride concentrations do not cause any damage. Corrosive effects may start to develop, however, if a critical chloride concentration has been reached in the concrete.

With regard to frost and deicing salt attack on concrete the chloride concentration in the near-surface concrete section controls the freezing behaviour of the pore solution in the capillary pores. Owing to dissolved hydration products the pore solutions in larger capillaries start to freeze only at temperatures below approximately –4°C, and further depression of the freezing point occurs with increasing concentration of chlorides: e.g. down to –21°C for a saturated sodium chloride solution [6.17].

Increasing concentration of the salt in the pore system also increases the moisture concentration due to the hygroscopic effect of the salt, and more freezable water is present in the pore system. As explained earlier, however, at the same time the freezing temperature is reduced. Also, secondary effects such as salt crystallization in the pore system upon drying in principle increase with increasing salt concentration.

These individual mechanisms gradually become more pronounced the higher the salt concentration, and therefore no threshold concentration exists.

As was stated in section 6.2.2 the corrosive action of chlorides on steel reinforcement consists mainly of the local depassivation of the steel surface causing pitting corrosion. Additionally, the free chloride ions in the pore solution gradually reduce the ohmic resistance of the concrete section between anodic and cathodic sites, thus encouraging high corrosion rates for pitting corrosion and for general corrosion as well [6.8, 6.22, 6.31]; see Fig. 6.3.

The drop in ohmic resistance increases gradually with increasing chloride concentration. Depassivation of the steel surface, however, requires a chloride concentration exceeding a threshold value, which is often referred to as the **critical chloride concentration**.

The critical chloride concentration is not a constant value for any type of concrete or exposure condition but depends on many parameters, the most important being the composition of the pore solution. From laboratory experiments on steel immersed in a solution of known pH it was derived that the breakdown of the passive layer is controlled by the free chloride ion concentration in relation to the hydroxyl ion concentration and a threshold value of $Cl^-/OH^- = 0.6$ has been reported. For steel embedded in concrete the conditions for passivation may be less favourable, and therefore lower threshold values may be encountered [6.22].

It is evident that the above-mentioned ratio depends on the type of cement as well as on the cement content used for the production of a concrete. Both parameters control the amount of free chloride ions in the pore solution due to differences in the binding capacity, and the hydroxyl ion concentration may vary, especially for blended cements or concrete

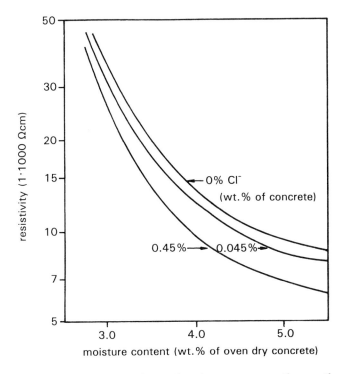

Fig. 6.3. Effect of free chloride ion and moisture concentration on the resistivity of concrete [6.31].

mixes with additions of pozzolans [6.22, 6.26, 6.32, 6.33]. Furthermore, it has been reported that the threshold concentration for chlorides may depend on the type of salt: i.e. calcium chloride, sodium chloride or potassium chloride [6.26].

The critical ratio of the relative free chloride ion concentration also explains the crucial effect of carbonation on chloride-induced corrosion. Decomposition of the hydration products liberates the combined chloride, thus increasing the free chloride ion concentration, while at the same time the OH^- concentration is reduced in the course of neutralization. For carbonated concrete the critical concentration is therefore reached already for very low total chloride levels.

The critical chloride concentration is an important parameter for estimating the lifetime of a reinforced concrete structure subjected to chlorides as well as for repair or rehabilitation of structures where chlorides have penetrated into the concrete cover but no reinforcement corrosion has yet taken place. In these cases a decision must be made on a permissible chloride content of the concrete that can be accepted without endangering the corrosion protection of the reinforcement. Because of

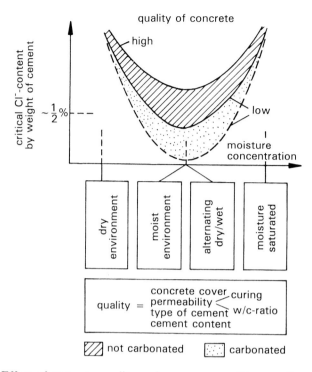

Fig. 6.4. Effect of concrete quality and exposure condition on the critical chloride concentration for reinforced concrete [6.6, 6.39].

differences in the binding capacity, a permissible chloride concentration will depend of course on the concrete composition, but also the perviousness of the concrete cover, the depth of carbonation, the oxygen supply and the moisture state must be considered. These additional parameters will affect the free chloride ion concentration, the electrode potential of the reinforcement and the ohmic resistance of the concrete cover. According to these considerations acceptable chloride concentrations, i.e. the critical chloride concentration, in most cases will range from 0 to approximately 1% chloride by mass of the cement: see Fig. 6.4 [6, 34].

Chloride corrosion is self-supporting; therefore the above-mentioned limits will apply only as long as no corrosion process has started yet.

6.5 Transport mechanisms for chlorides

For the ingress of chlorides into concrete the continuous network of the capillary pore system of the hcp, the coarse pore system of the aggregate/matrix interface and, eventually, microcracks provide the paths along which the transport of ions occurs. As mentioned earlier, different physical mechanisms may be distinguished with respect to the driving force, and in the terminology according to Chapter 2 these mechanisms may be referred to as the **permeation** of a salt solution, the **capillary absorption** of chloride-containing liquids, and the **diffusion** of free chloride ions. Depending on the exposure condition as well as on the moisture content of the concrete element these individual mechanisms may act simultaneously; they may prevail in sequence during consecutive periods of time; or one of them may be the exclusive transport mechanism. Similar considerations are valid with respect to the location within a concrete element: if a chloride solution penetrates through the near-surface layer by non-steady permeation, at the penetration front capillary forces become predominant along the coarse capillaries, and the diffusion of free ions then leads to a homogeneous distribution of the ions also into remote sites of the paste. In most cases, therefore, mixed modes of chloride transport prevail.

For the individual basic physical phenomena, rather simple empirical laws have been established for a mathematical description of the transport processes; see Chapter 2. Although they may serve for rough estimates of penetration rates, concentration profiles or the depth of penetration with time, a variety of necessary simplifications, the lack of reliable material parameters and the treatment of a non-coupled flow limit the precision of the corresponding prediction models.

6.5.1 PERMEATION

Permeation as a relevant transport mechanism may become important for the ingress of chlorides only if the penetrating liquid carries chloride ions, e.g. dissolved salts such as in seawater or from industrial processes. In this case the transport of chlorides is a convective flow, which is governed by the flow of the liquid phase and its chloride concentration.

The permeation of a fluid, in this case a liquid, is described by Hagen-Poiseuille according to eq. (2.8), Chapter 2, assuming a laminar and saturated capillary flow. If a salt solution is considered, the effect of the type of salt and its concentration on the viscosity must be taken into account. However, for low concentrations of chlorides in solution as they occur in nature this effect is small and may be neglected in most cases [6.35].

During an initial period of penetration, chlorides from the flowing salt solution will be combined by the hydration products of the cement until equilibrium conditions between free and combined chloride ions are achieved; see section 6.3. In principle, during this stage the concentration of the chloride solution decreases with increasing depth of penetration.

The permeation of a chloride solution may be a relevant transport mechanism for chloride ingress only in rare cases such as marine structures under a high hydrostatic pressure or for the seepage of solutions through retaining structures.

6.5.2 CAPILLARY SUCTION

Similar to permeation, the ingress of chlorides due to the capillary action of the pore system absorbing a chloride-containing solution is a convective flow of ions, and the amount of chloride introduced is given by the volume of absorbed solution and its chloride concentration. Equations (2.12) and (2.15) in Chapter 2 demonstrate that the driving force is controlled by the pore radius and the effective surface tension of the liquid, and the density as well as the viscosity of the liquid must be considered for the rate of penetration or the height of the capillary rise; see eq. (2.15), Chapter 2.

According to these considerations a reduction of the capillary suction should be expected for salt solutions with increasing salt concentration as compared to pure water. However, the observed suction behaviour for salt solution is discussed controversially by various workers.

In an experimental investigation, higher absorption rates have been observed for sodium and calcium chloride solutions than for pure water [6.36]; see Figs 6.5 and 6.6.

In contrast, no differences in the capillary absorption of dilute solutions of calcium chloride or sodium chloride could be observed by Volkwein [6.37]. He concluded that other factors controlling the suction of a concrete are more important than the change in surface tension, viscosity and

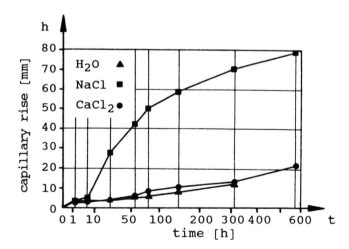

Fig. 6.5. Capillary rise of 2-molar solutions of sodium and calcium chloride and of pure water [6.36].

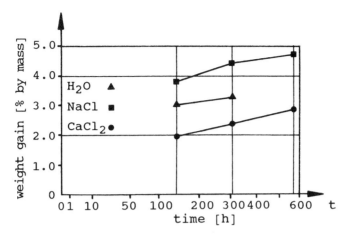

Fig. 6.6. Weight gain of mortar prisms due to capillary absorption of 2-molar solutions of sodium or calcium chloride and pure water [6.36].

density of the liquid. In his investigations it was shown that neither the absorption of pure water nor that of a salt solution obeys the $\sqrt{t}$ relation. Furthermore, the penetration front of the chlorides in most cases was lagging behind the water front. This filtration process was attributed to the binding of the chloride ions at the hydration products.

The absorption of chloride-containing solutions must be considered as the major cause of chloride contamination of concrete that is subject to alternating exposure conditions with drying/wetting cycles. During wetting periods the near-surface concrete layers readily absorb the chloride solution. Subsequently, the water will evaporate upon drying, whereas the salt remains in the pore system of the near-surface region. Thus, repeated drying/wetting cycles increase the salt concentration in the pore system, which may become even higher than the concentration of the outer chloride solution. Depending on the relative humidity of the environment the hygroscopicity of the salt prevents or at least lowers the evaporation of water from the concrete surface zone, thus increasing also the moisture concentration with increasing salt contamination. If sufficient liquid paths are provided the ions of the salt then penetrate deeper into the concrete, for instance by diffusion.

6.5.3 DIFFUSION

In contrast to the convective flow of permeation or capillary absorption of chloride-containing solutions, the diffusion of chloride ions into concrete is caused by gradients of the chloride concentration. Although this transport mechanism does not depend on the flow of water as a vehicle for chloride ions, water plays an essential role in the transport process. Only if a sufficient moisture concentration provides for continuous liquid paths in the capillary pore system can the chloride ions penetrate the matrix. No exact data are available regarding the diffusion rate as a function of the moisture concentration. However, as the transport occurs along the capillary pores these pore spaces should be filled with water or at least contain a high amount of water. The diffusion mechanism ceases if the liquid paths become interrupted due to drying. The highest diffusivity, therefore, should be expected for water-saturated material, and a continuous decrease should occur with decreasing moisture content. A moisture concentration in equilibrium with approximately 60–80% RH may be regarded as the lower limit allowing ion diffusion [6.38, 6.39]. It is worthwhile to note that this humidity range coincides with the upper limit for the free diffusion of gases such as oxygen or carbon dioxide, which depends on a continuous network of almost empty pores. Aside from pore structure characteristics the moisture concentration is thus a major parameter for the chloride ion diffusion. An estimate of the chloride ion diffusion coefficient from concentration profiles taken on one particular concrete but measured on an interior (dry) or exterior (humid) surface resulted in apparent differences of almost one order of magnitude [6.40].

Another environmental parameter influencing ion diffusion is given by the temperature [6.40, 6.42], and the diffusion coefficient D may be written as a function of the temperature according to the Arrhenius equation:

$$D(T) = D_0 \times e^{-B/RT}$$

where
D_0, R, B = constants
T = absolute temperature
$D(T)$ = diffusion coefficient at temperature T

Within a certain range of temperature T, this equation postulates a progressive increase of the diffusion coefficient with rising temperature. Whiting [6.40] has summarized data available from the literature that support this model; see Table 6.1.

Temperature may also affect the binding of chloride ions by the hydration products and thus additionally influence the depth of penetration into concrete.

In the diffusion of ions the electric charge transported must also be considered. In order to maintain neutrality the negative charge carried by the penetrating chloride ions must be balanced by a corresponding flow of other ions, i.e. positive ions simultaneously moving into the concrete or a counter-diffusion of negative ions [6.13], or balance must be achieved by an ion exchange with hydration products [6.37]. These processes explain that the rate of chloride penetration also depends on the cation involved

Table 6.1. Effect of temperature upon chloride diffusion coefficients

Author	w/c	Temp. (°C)	Curing	D_e $(m^2/s \times 10^{-12})$	Type of mix
Collepardi (1972)	0.40	10	2 mo mist 1 mo air	1.2	Paste
		25		2.5	"
		40		4.9	"
	0.40	25	"	1.7	Concrete
Goto (1981)	0.40	35	4 wk at 60°C	2.7	Paste
		45	"	3.7	"
		60	"	9.2	"
Page (1981)	0.40	7	60 d mist	1.1	Paste
		25	"	2.6	"
		44	"	8.4	"
	0.50	7	"	2.1	"
		25	"	4.5	"
		44	"	18.4	"
	0.60	7	"	5.2	"
		25	"	12.3	"
		44	"	31.8	"
Brodersen (1982) [6.29]	0.6	15	7 d mist	4.21	Paste
		21	"	6.21	"
		25	"	7.83	"
		30	"	10.79	"

as well as on the composition of the concrete pore solution, which may change drastically due to concrete additions or blending of the cement clinker with pozzolans or slag [6.41]. It also explains that binding as well as the diffusion rate are different for various chloride-bearing salts. It has been reported that the diffusion coefficients for the chloride ion increase in the following order [6.42]:

$$NaCl - CaCl_2 - MgCl_2$$

The theoretical description of the chloride ion ingress due to a diffusion process is basically done according to Fick's second law of diffusion. However, a major complication arises from the binding of chloride ions by the hydration products. Chlorides penetrating into an element are combined partially until equilibrium with the pore solution is achieved and as long as the binding capability of the paste is not yet exhausted. In the balance equation the binding is then taken into account by the additional term representing a sink; see eq. (2.5), Chapter 2. This approach, however, requires detailed information on parameters describing the binding process, such as the binding capacity for a given concrete mix, the influence of temperature and moisture concentration, and the time rate of binding. Sufficient data are not yet available on this matter for a comprehensive prediction model. A theoretical approach considering diffusion as the relevant transport process and binding of chlorides in chemical reactions with hcp compounds with the help of a Langmuir sorption isotherm has been presented (see ref. 147 in [6.28]).

As a driving force only a gradient of the free chloride ions in the pore solution is effective, and the diffusion coefficient refers to the diffusion of free ions.

A different approach is often applied if diffusion coefficients are derived from measured concentration profiles or if observed concentration profiles in existing structures serve as a basis for estimates on the future progress of the chloride contamination. No distinction is then made between free and combined chlorides; the concentration refers to the total chloride content held in a volume element, and an effective diffusion coefficient D_{eff} is introduced. This method does not describe the true processes involved in the transport mechanisms, and therefore the validity of the approach is limited.

6.5.4 MIXED MODES

In the preceeding sections the ingress of chlorides has been discussed on the basis of different but distinctly separate transport mechanisms. The presence of such conditions solely depends on appropriate boundary conditions as well as on the moisture state and its distribution in the concrete element considered. Pure permeation of a chloride solution as well as pure diffusion of chloride ions will prevail only for a moisture-saturated concrete in which no capillary forces can be active. If dry or non-saturated concrete is exposed to a chloride solution, however, capillary absorption is the dominant mechanism. Nevertheless, small hydraulic pressure heads can support the ingress by permeation and the diffusion of ions simultaneously carries the ions also into narrow pore spaces where no capillary flow occurs anymore. Except for concrete elements that are continuously submerged in seawater these mixed modes of the chloride transport obviously prevail in most cases for structures in service.

For a theoretical treatment of these mixed modes the individual transport mechanisms must be superimposed, thereby considering the individual driving forces and the relevant transport parameters. Among the various factors influencing these transport parameters the local moisture concentration deserves special attention, because the moisture concentration does not only represent the most important influencing parameter on the transport coefficients. Gradients with respect to moisture cause a moisture flow, which also carries ions. Therefore, a coupled system of differential equations for the moisture flow in the vapour phase, non-saturated and saturated capillary flow as well as ion flow must be modelled in which their mutual dependences are considered: for example the effect of moisture on the transport of ions as well as a moisture flow due to a gradient in ion concentration, i.e. osmosis.

Although a numerical solution of such models can be obtained, e.g. by finite elements, the required material parameters are not available at present.

6.6 Chloride profiles

The spatial distribution of the chlorides in concrete elements is a function of the exposure conditions and time. For cases where constant boundary conditions prevail with respect to moisture and chloride concentration and diffusion theory may be applicable, concentration profiles according to Fig. 6.7 are obtained as a result of Fick's second law of diffusion [6.43]; see eq. (2.3), Chapter 2. In this diagram the parameter Dt/l^2 (Fourier number) shows that the penetration depth increases with increasing duration of exposure t and increasing diffusion coefficient D.

Similar concentration profiles are also often observed in structures in service as well as in laboratory experiments in which concrete specimens have been immersed in chloride solutions. As discussed further in Chapter 9, such concentration profiles are then used to calculate the diffusion

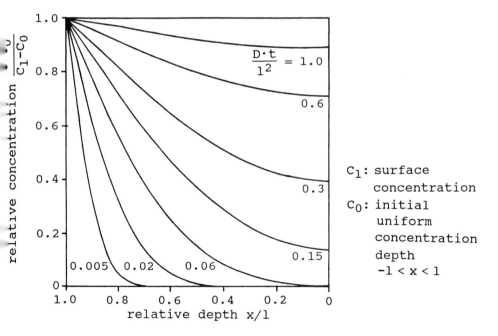

Fig. 6.7. Concentration profiles as a result of Fick's second law of diffusion [6.43].

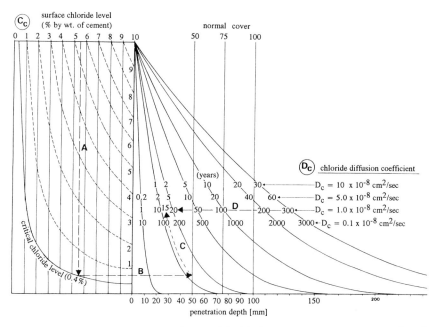

Fig. 6.8 Nomogram for penetration of chloride into concrete according to [6.31].

coefficient for chlorides in concrete, or to estimate the progress of the chloride penetration with time; see Fig. 6.8 [6.31].

Whereas Fig. 6.7 presents solutions of Fick's second law in the most general way, thereby introducing dimensionless parameters, i.e. a relative concentration of $(c - c_0)/(c_1 - c_0)$, a relative depth of penetration x/l, and the time as given by the Fourier number Dt/l^2, Fig. 6.8 gives results of Fick's law for absolute values of a surface concentration c_1, the depth of penetration x, as well as combinations of diffusion coefficients D with time t.

In this approach it must be kept in mind that an exclusive diffusion process was assumed; furthermore, problems associated with chloride binding are encountered.

Varying boundary conditions as well as combined transport mechanisms will result in deviations of the chloride profiles from those given in Figs 6.12 and 6.13. Wash-out of chlorides from the near-surface regions can occur if the surface of the concrete element comes into contact with a solution of low concentration or pure water. Then the maximum chloride concentration of the chlorides no longer prevails at the surface zone but in the inner section at some distance from the surface; see Fig. 6.9 [6.10].

High concentrations of chlorides may be achieved in locations where sustained or repeated evaporation of the pore solution containing chlorides occurs. Aside from repeated cycles of drying and wetting with a chloride

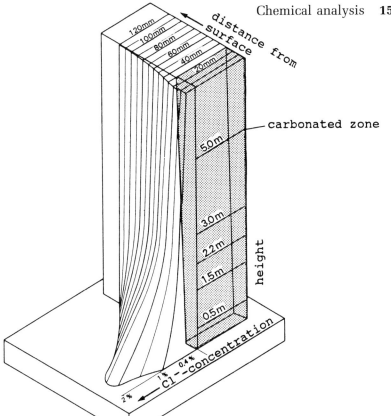

Fig. 6.9. Chloride distribution in a concrete element subjected to deicing salt solution in the splash/spray zone according to [6.10].

solution, such conditions are encountered for partially immersed structures where the take-up of the chloride solution is due to capillary absorption, and evaporation of the water can take place at some distance from the water level; see Fig. 6.10 [6.42].

In contrast to these results, Gjorv [6.44] found higher concentrations in the continuously submerged zones of concrete structures. Upon demolition of an 80-year-old marine structure, samples were taken from concrete pillars and the chloride concentration profiles were analysed. Continuously submerged parts of the pillars always exhibited higher chloride concentrations than the concrete taken from the tidal zone or upper sections of the members.

6.7 Chemical analysis

Damage evaluation, estimates of the expected lifetime and the selection of appropriate measures for the maintenance as well as for repair of

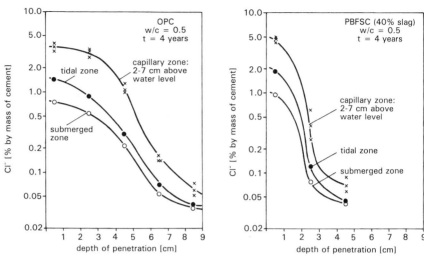

Fig. 6.10. Chloride concentration profiles observed in concrete elements partially immersed in seawater [6.42].

reinforced concrete structures are common engineering tasks, which often also require information on the chloride contamination of the concrete and their distribution across a concrete section. Concentration profiles for chlorides are also required for the experimental determination of transport coefficients, for instance by immersion tests (see Chapter 9). The experimental determination of concentration profiles thereby involves two major problems, which relate to sampling and to the chemical analysis of the sampled material with respect to its chloride content.

6.7.1 SAMPLING

The depth of penetration of chlorides into a concrete section is usually limited to a depth from several millimetres to a few centimetres only, thereby exhibiting steep concentration gradients in the near-surface sections. If a realistic concentration profile must be established from experimental analyses, samples must be taken from the concrete element at different depths from the surface. A high spatial resolution is required if special effects are to be detected, such as surface wash-out, or if diffusion coefficients are to be derived from concentration profiles. The thickness of the analysed section should therefore be small. However, concrete is a heterogeneous material, and with increasing size of the aggregates small samples become less representative for a given section.

Care must be taken in gaining the test samples from the concrete element in order not to affect the chloride content. Although taking samples by crushing will not affect the chloride content to be analysed this method in most cases will be inappropriate because the material gained may not be referred to an exact depth.

Drilling cores that are subsequently cut into thin slices allows for an exact determination of the sampling depth. This procedure however bears the risk of changing the chloride content of the samples in the process of drilling and cutting. The risk can be reduced if dry cutting and drilling is applied. If water cooling is required in drilling then bigger core diameters should be chosen, and a minimum diameter of the core of 100 mm has been recommended. In contrast, drilled cores of 30–50 mm in diameter are recommended for dry cutting procedures [6.44].

Another method that has been widely used in recent years consists of collecting drill powder or the powder that results from local grinding. No water is applied and, therefore, the risk of changing the chloride concentration in the test material is minimized. The method also allows an exact determination of the sampling depth. The disadvantage of this method is the very small amount of material available for analysis. Considering the heterogeneous structure of concrete a significant scatter of the test results should be expected, and therefore several samples may be required to calculate a reliable average chloride concentration for a certain sampling depth.

Table 6.2 gives the minimum number of drill holes recommended, thereby considering the diameter of the drilling tool as well as the maximum aggregate size of the concrete [44].

The amount of powder also depends on the thickness of the concrete section chosen for analysis. The values given in Table 6.2 are based on a thickness of approximately 10 mm for each individual sampling depth. If a higher resolution is desired the number of samples has to be increased correspondingly.

Although the ingress of chlorides into concrete for normal exposure conditions is usually limited to several centimetres only, it is recommended to take samples for chloride analysis also from deep sections of the interior part of the concrete member for chloride analysis, because a certain level of chloride contamination of the concrete will always have been

Table 6.2. Number of drill holes for collecting concrete drill powder [6.45]

Max. aggregate size (mm)	Number of drill holes with diameters (mm)			
	20	26	32	40
8	1	1	1	1
16	2	1	1	1
32	5	3	2	1

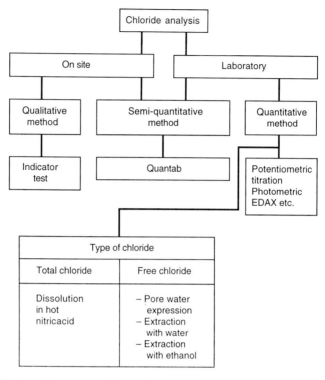

Fig. 6.11. Principles of cholride analysis.

caused already by the chlorides contained in the raw materials for concrete production.

6.7.2 ANALYSIS OF THE CHLORIDE CONTENT

Different methods are available for the measurement of the chloride contamination of concrete, on either a mere qualitative or fully quantitative basis. The selection of a particular method will depend primarily on the purpose of the analysis with regard to the required exactness, other factors being the available equipment, time considerations etc. In any case the analysis of chloride concentration in concrete should be conducted only by trained and experienced staff. Figure 6.11 gives an overview of the available methods and their applications.

Similar to the phenolphthalein test for the rapid determination of the carbonation depth, an indicator test is available for chlorides in concrete. Applied to a fresh fracture plane this test can easily be conducted on site, and it provides rapid information on the presence of chlorides. This test may even be of help if exact laboratory analyses are intended later on.

Recently, a test kit has been introduced that comprises a grinding device for sampling concrete powder as well as a test kit for the analysis of the total acid-soluble chloride content of the powder. The kit is designed to fully operate on site, and test results are reported to be available within 2 h [6.49]. The results of the rapid chloride test (RCT) have been correlated to standard laboratory methods.

Except for energy-dispersive X-ray analyses the semi-quantitative and the quantitative methods require grinding of the concrete sample and subsequent extraction of the chlorides. For analysis of the total chloride content the sample is usually dissolved in hot nitric acid [6.45–6.47].

Especially with regard to the corrosion protection of reinforcement or for the evaluation of the chloride binding capacity of different cements a distinction between combined and free chlorides or total chloride content and concentration of free chloride ions in pore water is necessary. For extraction of the free chloride ion content, different extraction methods have been proposed such as pore water expression or washing out of the chlorides with either water or ethanol alcohol. However, much uncertainty still remains with the latter methods for the determination of the free chloride ion content if and to what extent combined ions are also extracted [6.9, 6.12, 6.33, 6.45].

The amount of chloride ions determined in the sample is usually referred to the mass of concrete. Often more important, however, is the chloride concentration in relation to the cement content. If the cement content of the concrete is unknown, a corresponding chemical analysis should be conducted.

6.8 References

6.1. ENV 206 (1990) Concrete – Performance, production, placing and compliance criteria.
6.2. Comité Euro-International du Beton – CEB (1991) *CEB-FIP Model Code 1990*, Bulletin d'Information No. 203, 204, 205, Lausanne.
6.3. RILEM CPC 14-79, *Concrete test methods – Mixing water for concrete.*
6.4. Chlorid im Beton Sachstandbericht des Arbeitskreis 'Chlorid' des Deutschen Betonvereins, Fassung Dezember 1982 in 'DBV-Merkblattsammlung', Deutscher Betonverein e.V. Wiesbaden, 1991
6.5. Mehta, P.K. (1980) *Durability of concrete in marine environment – A review*, ACI SP 65-1, Detroit.
6.6. Comité Euro-International du Beton – CEB (1989) *Durable concrete structures*, CEB Design Guide, Bulletin d'Information No. 182, 2nd Edition.
6.7. CEB–RILEM International Workshop (1983) Durability of Concrete Structures, Copenhagen, Workshop Report, Ed. S. Rostam.
6.8. Haynes, H. (1980) *Permeability of concrete in sea water*, ACI-SP 62-2, Detroit.

6.9. Volkwein, A. (1983) Beiträge zum Internationalen Kolloquium 'Chloridkorrosion'. *Mitteilungen aus dem Forschungsinstitut des Vereins der Österreichischen Zementfabrikanten*, Heft 36, Wien.

6.10. Weber, D. (1983) Untersuchungen zur Einwirkung von Tausalzen sowie zur Carbonatisierung an Bauwerken der Berliner Stadtautobahn, Int. Colloquium Materials Science and Restoration, Esslingen, Germany.

6.11. Schießl, P. Chlorid-Eindringtiefen – Rechen- und Beurteilungsmodell. *Forschungsberichte: Spätschäden an Spannbetonbauteilen* No. 45/20, Teil A + B, Institut für Betonstahl und Stahlbetonbau e.V., München.

6.12. Tuutti, K. (1983) Analysis of pore solution squeezed out of cement paste and mortar, Internationales Kolloquium 'Chlorid Korrosion', *Mitteilungen aus dem Forschungsinstitut des Vereins der Österreichischen Zementfabrikanten*, Heft 36, Wien.

6.13. Nilsson, L.O. and Tang Luping (1990) Chloride distributions in concrete, Draft report submitted to RILEM TC 116.

6.14. Wesche, K., Neroth, G. and Weber, J.W. (1982) Eindringen von Chlorid-Ionen aus PVC-Abbrand in Stahlbetonbauteile, *Deutscher Ausschuß für Stahlbeton*, Heft 352, Verlag Ernst & Sohn, Berlin.

6.15. Schießl, P. (1983) Maßnahmen bei Chloridschäden – PVC Brandgase Internationales Kolloquium 'Chlorid Korrosion', *Mitteilungen aus dem Forschungsinstitut des Vereins der Österreichischen Zementfabrikanten*, Heft 36, Wien.

6.16. Powers, T.C. (1958) The physical structure and engineering properties of concrete. *PCA Bulletin*, No. 90, Chicago.

6.17. Setzer, M.J. (1977) Einfluß des Wassergehaltes auf die Eigenschaften des erhärteten Betons, *Deutscher Ausschuß für Stahlbeton*, Heft 280, Verlag Ernst & Sohn, Berlin.

6.18. Fagerlund, G. (1975) The significance of critical degrees of saturation at freezing of porous and brittle materials, in *Durability of Concrete*, ACI SP-47, Detroit.

6.19. Mindess, S. and Young, J.F. (1981) *Concrete*, Prentice-Hall, Inc., Englewood Cliffs.

6.20. Günter, M., Bier, Th. and Hilsdorf, H. (1987) Effect of curing and type of cement on the resistance of concrete to freezing in deicing salt solutions, in *Concrete Durability – Katharine and Bryant Mather International Conference*, ACI SP-100, Detroit.

6.21. Hjorth, L. (1983) Development and application of high density cement based materials, in *Proc. Technology in the 1990s: Developments in hydraulic cements*, The Royal Society, London.

6.22. Schießl, P. (Ed.) (1988) *Corrosion of Steel in Concrete*, Report of RILEM Report Technical Committee 60-CSC, Chapman & Hall, London.

6.23. Uhlig, H.H. (1975) *Korrosion und Korrosionsschutz*, Akademie-Verlag, Berlin.

6.24. Treadaway, K.W.J., Brown, B.L. and Cox, R.N. (1978) Durability of galvanized steel in concrete, in *Corrosion of reinforcing steel in concrete*, ASTM STP 713, Philadelphia.

6.25. Tritthart, J. (1984) Bewehrungskorrosion – Zur Frage des Chloridbindevermögens von Zement. *Zement-Kalk-Gips*, Vol. 37, pp. 200-4.

6.26. Hannson, C.M. and Berke, N.S. (1988) Chlorides in concrete, in *Proc. Pore Structure and Permeability of Cementitious Materials*, Materials Research Society Symposium Proceedings, Vol. 137, Boston.

6.27. Regourd, M. (1980) Physico-chemical studies of cement pastes, mortars, and concretes exposed to sea water, in *Performance of concrete in marine environment*, ACI SP-65, Detroit.

6.28. Roy, D.M. (1986) Mechanisms of cement paste degradation due to chemical and physical factors, in *Proc. 8th Int. Congress on the Chemistry of Cements*, Rio de Janeiro.

6.29. Brodersen, H. (1982) Zur Abhängigkeit der Transportvorgänge verschiedener Ionen im Beton von Struktur und Zusammensetzung des Zementsteins. Dissertation, Technische Hochschule Aachen.

6.30. Page, C.L., Lambert, P. and Vassie, P.R.W. (1991) Investigations of reinforcement corrosion *Materials and Structures*, Vol. 24, pp. 243–252.

6.31. Browne, R.D. (1982) Design prediction of the life for reinforced concrete in marine and other chloride environments. *Durability of Building Materials*, Vol. 1, pp. 113–125.

6.32. Weigler, H. (1983) Einflüsse des Betons und seiner Bestandteile auf Chlorideinbindung und -diffusion in 'Chloridkorrosion'. *Mitteilungen aus dem Forschungsinstitut des Vereins der Österreichischen Zementfabrikanten*, Heft 36.

6.33. Page, C.L. and Vennesland, O. (1983) Pore solution composition and chloride binding capacity of silica-fume cement pastes. *Materials and Structures*, Vol. 16, pp. 19–25.

6.34. Grube, H., Kern, E. and Quitmann, H.P. (1990) Instandhaltung von Betonbauwerken, *Betonkalender 1990*, Ernst & Sohn.

6.35. Horvath, A.L. (1985) *Handbook of aqueous electrolyte solutions, physical properties, estimation and correlation methods*.

6.36. Wenger, B. (1978) Aufsaugversuche an Mörtelprismen mit Wasser und Tausalzlösungen *Beton*, Vol. 2, pp. 52–54.

6.37. Volkwein, A. (1991) Untersuchungen über das Eindringen von Wasser und Chlorid in Beton. *Berichte aus dem Baustoffinstitut*, Technische Universität München, Heft 1.

6.38. Großkurth, K.P. and Malorny, W. (1983) Chemische Einwirkung korrosiver Brandgasbestandteile auf Stahlbeton, Int. Kolloquium 'Chloridkorrosion', *Mitteilungen aus dem Forschungsinstitut der Österreichischen Zementfabrikanten*, Heft 36, Wien.

6.39. Locher, F.W. and Sprung, S. (1970) Einwirkung von salzsäurehaligen PVC-Brandgasen auf Beton. *Beton*, Vol. 20, No. 2, pp. 63–65, and No. 3, pp. 99–104.

6.40. Whiting, D. Private communication with TC 116.

6.41. Gjørv, O.E. and Vennesland, Ø. (1979) Diffusion of chloride ions from seawater into concrete. *Cement and Concrete Research*, Vol. 9, pp. 229–239.

6.42. Rehm, G., Nürnberger, U., Neubert, B. and Nenninger, F. (1988) Einfluß von Betongüte, Wasserhaushalt und Zeit auf das Eindringen von Chloriden in Beton, *Deutscher Ausschuß für Stahlbeton*, Heft 390, Beuth Verlag, Berlin.

6.43. Crank, J. (1970) *The Mathematics of Diffusion*, Oxford University Press.

6.44. Gjorv, O.E. and Kashino, N. (1986) Durability of a 60-year old reinforced concrete pier in Oslo Harbour. *Materials Performance*, Vol. 25, No. 2, pp. 18–26.

6.45. Deutscher Ausschuß für Stahlbeton (1989) Anleitung zur Bestimmung des Chloridgehaltes von Beton, Arbeitskreis 'Prüfverfahren Chlorid-eindringtiefe' des Deutschen Ausschusses für Stahlbeton; Heft 401, Beuth Verlag, Berlin.

6.46. Dorner, H. and Kleiner, G. (1989) Schnellbestimmung des Chloridgehaltes von Beton, *Deutscher Ausschuß für Stahlbeton*, Heft 401, Beuth Verlag, Berlin.

6.47. Dorner, H. (1989) Bestimmung des Chloridgehalts von Beton durch Direktpotentiometrie, *Deutscher Ausschuß für Stahlbeton*, Heft 401, Beuth Verlag, Berlin.

6.48. Smolczyk, H.G. (1968) Chemical reactions of strong chloride solutions with concrete, in *Proc. Fifth International Symposium on the Chemistry of Cement*, Tokyo, Supplementary paper III-31.

6.49. Petersen, C.G. (1993) RCT profile grinding kit for in-situ evaluation of the chloride diffusion coefficient and the remaining service life of a reinforced concrete structure, in *Chloride Penetration into Concrete Structures*, Ed. L.O. Nilsson, Chalmers Technical University, Göteborg, Publication P-93:1.

7 Concrete compressive strength, transport characteristics and durability

Hubert K. Hilsdorf

7.1 The problem

In engineering practice, codes and specifications, the standard compressive strength of concrete is used not only as a basis of structural design and as a criterion of structural performance, but also as a criterion for the durability of a concrete structure. Such an approach can be justified by the observation that both strength and transport characteristics are to a large extent linked to the pore structure of the concrete: low porosity results in high strength and also in a high resistance to the penetration of aggressive media.

In contrast to most transport characteristics, compressive strength can be determined on the basis of well-established and generally accepted routine test methods. In addition, the entire design process of a structure would be substantially simplified if one and the same material characteristic, i.e. a standard strength value, could be used both for structural design and for durability design.

In order to evaluate the validity and/or the limitations of such an approach some data from the literature are summarized in the following, from which correlations between strength and transport characteristics as well as between strength and durability characteristics can be deduced.

It is well known that the compressive strength of a concrete specimen of a given composition depends on a number of parameters, such as age, curing conditions, moisture state at the time of testing and size and shape of the test specimen. Furthermore, in the definition of a standard strength, distinction is made between an average strength and a characteristic strength that corresponds to a certain cut-off or fractile value. In correlating strength data, transport characteristics and durability data it is, therefore, essential to define the conditions for which these data had been obtained.

Performance Criteria for Concrete Durability, edited by J. Kropp and H.K. Hilsdorf.
Published in 1995 by E & FN Spon, London. ISBN 0 419 19880 6.

7.2   Compressive strength and transport characteristics

7.2.1   PERMEABILITY

(a) General considerations

According to Powers *et al.* [7.1, 7.2] the compressive strength of a hydrated cement paste can be correlated with the capillary porosity of the paste by

$$f_{cp} = f_{cpo}\left(1 - v_{cp}\right)^{n}$$

(7.1)

where f_{cp} = compressive strength of the paste
 f_{cpo} = compressive strength of a paste free of capillary pores
 v_{cp} = capillary porosity expressed as a fraction of the total paste volume minus the volume of unhydrated cement
 n = exponent

Figure 7.1 shows experimental data on the water permeability of hydrated cement pastes from [7.3]. The water permeability coefficient of mature pastes is given as a function of the water/cement ratio of the pastes. From the information given in [7.1] and [7.2], the capillary porosity of the paste can be estimated from the water/cement ratio under the assumption of a certain degree of hydration. On the basis of such an evaluation

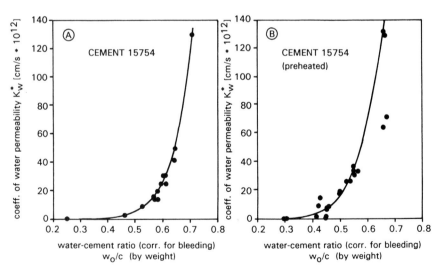

Fig. 7.1. Relationships between coefficient of water permeability and water/cement ratio for mature paste [7.3].

the permeability coefficients given in Fig. 7.1 also can be presented as a function of capillary porosity v_{cp} as defined above. This relation can be expressed very well by

$$K_{wp} = \frac{K_{wpo}}{\left(1 - v_{cp}\right)^m}$$

$$(7.2)$$

where K_{wp} = coefficient of water permeability of the paste
$\quad\quad\ K_{wpo}$ = coefficient of water permeability of a paste free of capillary pores
$\quad\quad\ m$ = exponent

Other formulations have been used to correlate capillary porosity and strength. In addition eqs (7.1) and (7.2) do not take into account the important influence of pore size distribution. Nevertheless, from eqs (7.1) and (7.2) a correlation between compressive strength and water permeability can be deduced, which should be valid at least as a rough approximation (eq. 7.3):

$$K_{wp} = K_{wpo}\left(\frac{f_{cpo}}{f_{cp}}\right)^{m/n}$$

$$(7.3)$$

When considering gas permeability the total capillary porosity of the paste v_{cp} should be replaced in eq. (7.2) by the empty capillary porosity v_{cpe}, i.e. that fraction of the capillary pore volume that is not filled with water. Setting

$$v_{cpe} = \alpha \cdot v_{cp}$$

we obtain, for the gas permeability of the paste, K_{gp}:

$$K_{gp} = \frac{K_{gpo}}{\left(1 - \alpha \cdot v_{cp}\right)^r}$$

$$(7.4)$$

where K_{gpo} = coefficient of gas permeability of the paste without empty pores
$\quad\quad\ r$ = exponent

From eqs (7.1) and (7.4) we obtain a correlation between compressive strength of the paste f_{cp} and gas permeability of the paste K_{gp}:

$$K_{gp} = \frac{K_{gpo}}{\left\{ 1 - \alpha \left[1 - \left(\dfrac{f_{cp}}{f_{cpo}} \right)^{1/n} \right] \right\}^r}$$

(7.5)

For $\alpha = 1$, i.e. all capillary pores are empty, eq. (7.5) yields eq. (7.6), which is similar to eq. (7.3). For $= 0$, i.e. all capillary pores are water-filled, we obtain $K_{gp} = K_{gpo}$.

$$K_{gp} = K_{gpo} \left(\frac{f_{cpo}}{f_{cp}} \right)^{r/n}$$

(7.6)

Equations (7.1)–(7.6) are valid for hydrated cement paste. For concrete the effect of the paste–aggregate interfaces and of the permeability of the aggregate should be taken into account so that

$$K_{gc} = \beta \cdot k_{wp}$$

and

$$K_{gc} = \gamma \cdot K_{gp}$$

where the coefficients and are not necessarily constant and may depend on paste content and maximum size and type of aggregates; refer to section 4.1.

(b) Experimental data

In Fig. 7.2 experimental data from [7.4] are presented. Oxygen permeability of concretes made with different water/cement ratios and cement contents as well as four types of cement is given as a function of the average compressive strength, which was determined as equivalent cube strength from the remaining halves of flexure specimens that had been moist-cured up to an age of 28 days. Oxygen permeability was measured on circular discs with diameter 150 mm and thickness 50 mm, which had been obtained from cores taken from the bottom part of a slab. The slabs were moist-cured for 7 days. Prior to permeability testing at an age of 28 days, the discs were preconditioned by oven drying them for 6 days at 50°C. According to Fig. 7.2 from [7.4], gas permeability and standard compressive strength as defined above can be correlated according to:

$$K_g = 7.5 \exp\left(-0.06 f_{cm\,28}\right)$$

(7.7)

where K_g =   oxygen permeability (10^{-16} m^2)
        f_{cm28} =   average compressive strength at an age of 28 days (MPa)

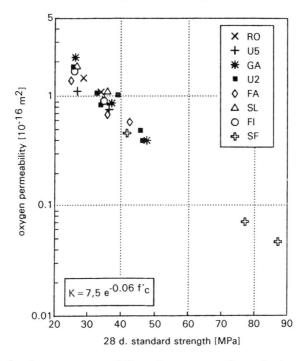

Fig. 7.2. Relation between permeability of concrete and standard strength for concretes made of different types of cement [7.4].

From the data given in [7.4] it follows that this correlation is independent of concrete mix proportions and type of cement. According to Fig. 7.2 and eq. (7.7) an increase of the compressive strength from 25 MPa to 50 MPa results in a reduction of gas permeability K_g from $\sim 2 \times 10^{-16}$ m^2 to $\sim 0.4 \times 10^{-16}$ m^2, i.e. $K_{g25} \sim 5 \times K_{g50}$.

Similar data from [7.5] and [7.6] are presented in Figs 7.3 and 7.4. There, the coefficient of air permeability of concretes made of different water/cement ratios, cement contents and types of cement, moist-cured for either 1, 3 or 7 days, is given as a function of the average compressive strength. The strength data had been obtained from 150 mm cubes, moist-cured for 7 days and tested at an age of 28 days. The air permeability was determined on circular discs diameter 150 mm, thickness 40 mm. After curing, the discs were stored in a constant environment of 65% RH and $T = 20°$C, up to the time of testing at an age of 56 days without further preconditioning. In Fig. 7.3 distinction is made between concretes made of different types of cement, whereas in Fig. 7.4 the duration of curing is used as an additional parameter. From Fig. 7.3 it follows that for a given compressive strength the air permeability coefficient may vary by almost 2 orders of magnitude; however, the type of cement does

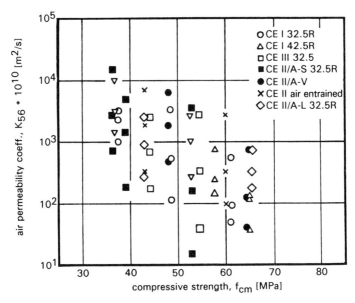

Fig. 7.3. Air permeability of concrete and compressive strength for concretes made of different types of cement and cured for 1, 3 or 7 days [7.6].

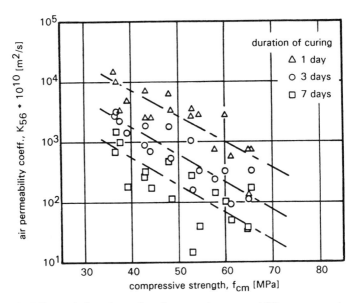

Fig. 7.4. Effect of duration of curing on air permeability–compressive strength relationships [7.6].

not seem to be a major parameter. From Fig. 7.4 it may be concluded that different air permeability compressive strength relations exist, which depend on the duration of curing of the permeability specimens. The correlation between air permeability and average compressive strength can be described by

$$K_g = K_{go} \cdot 10^{-b \cdot f_{cm\,28}/f_{cmo}}$$

(7.8)

where K_g = air permeability (10^{-16} m^2/s)
f_{cm28} = average compressive strength at an age of 28 days
f_{cmo} = 1 MPa
K_{go} = coefficient (10^{-16} m^2/s)

For a duration of curing of 7 days, values of K_{go} = 1.7 x 10^{-6} m^2/s and b = 0.044 have been obtained from a least-squares fit. Though eqs (7.7) and (7.8) are similar in their general form, the influence of compressive strength on gas permeability deduced from eq. (7.8) or Fig. 7.4 is much more pronounced than estimated from eq. (7.7). According to Fig. 7.4 an increase of the compressive strength, f_{cm28}, from 25 MPa to 50 MPa results in a reduction of air permeability K_g for 7 days of curing by a factor of approximately 15, whereas a factor of approximately 5 is obtained from eq. (7.7).

From the semi-theoretical considerations for gas permeability and strength expressed by eqs (7.5) and (7.6) a linear correlation between the logarithm of K_g and the logarithm of f_{cm28} should be expected for values of α approaching unity, whereas eqs (7.7) and (7.8) describe a linear relation between the logarithm of K_g and the actual values of f_{cm}. An evaluation of the data from [7.4] on the basis of log K_g = f(log f_{cm28}) results in a linear function with a correlation coefficient of 0.96, which is about the same as the one obtained for log K_g = $f(f_{cm28})$. The same holds true for the data from [7.5] and [7.6], so that no conclusions with respect to the general validity of eqs (7.5) and (7.6) can be drawn from these data.

A comparison of the data from [7.4] (Fig. 7.2) with those from [7.5] and [7.6] (Figs 7.3 and 7.4) allows some general conclusions:

● The correlation between permeability and 'standard' compressive strength is unique only if both parameters reflect the same pore structure of the concrete. If curing of the permeability specimens deviates from that of the strength specimens, relations between permeability and standard strength are obtained that depend on the particular curing regimes. It is likely, though unproven, that even for different curing regimes a unique strength permeability relation exists if both the strength and the permeability specimens are subjected to the same curing and tested at the same age.

- The correlations of strength and permeability data depend strongly on the preconditioning of the specimens used for the gas permeability measurements. The permeability specimens from [7.4] were oven-dried for 6 days at 50°C so that a uniform moisture state has been approached, resulting in a more or less constant and high ratio of empty to total capillary porosity. The specimens in [7.5] and [7.6] were only air-dried at room temperature up to an age of 56 days. Consequently, the denser concretes of higher strength had, because of their slower rate of drying, lower relative empty porosities than the lower-strength concretes. As a consequence, the permeability of the higher-strength concretes was underestimated, resulting in the higher ratio K_{g25}/K_{g50}, which can be deduced from the data in Figs 7.3 and 7.4 compared with the ratio K_{g25}/K_{g50} that follows from the data in Fig. 7.2.

In order to overcome some of the difficulties described above it has been proposed in [7.7] and [7.8] to use the strength of concrete at the end of curing as a durability parameter. In Fig. 7.5 from [7.7] the air permeability of the outer layer, thickness 35 mm, of concrete cubes, is given as a function of the cube strength of the concrete at the end of curing, f_{cmc}. These permeability values have been determined by means of a test method that is described in section 10.2.6. Concretes made of different types of cement and moist-cured between 28 and 360 days have been investigated. Prior to the measurement of air permeability the concrete has been air-dried for 80 days. From Fig. 7.5 it follows that there is a unique relation between

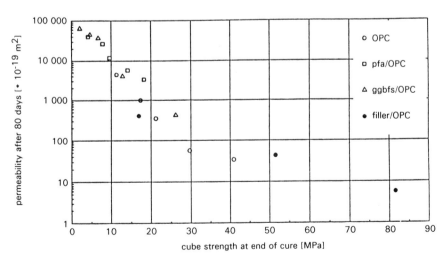

Fig. 7.5. Air permeability after 80 days drying versus cube strength at end of curing period [7.7].

air permeability K_g and concrete compressive strength at the end of curing (f_{cmc}), which is independent of duration of curing and type of cement. These data can be represented well by a linear function, log $K_g = f$ (log f_{cmc}), which corresponds to the general form of eq. (7.6).

A comparison of the data presented in Fig. 7.5 with those from Fig. 7.2 is noteworthy. For low values of compressive strength, the air permeabilities obtained from Figs 7.2 and 7.5 are almost identical, whereas for higher-strength concretes the coefficients of air permeability from Fig. 7.5 are lower than those obtained from Fig. 7.2. Thus from Fig. 7.5 a ratio of $K_{g25}/K_{g50} \sim 9$ is obtained compared with a value of $K_{g25}/K_{g50} \sim 5$ for the data from Fig. 7.2. This may be due to differences in the moisture states of the concretes at the time of permeability testing: the specimens from which the data of Fig. 7.2 have been obtained were oven-dried prior to permeability testing, whereas the specimens for the data from Fig. 7.5 had only been air-dried. This is of particular significance for the concretes of higher strength, which dry at a slower rate.

The uniqueness of the relation between permeability and compressive strength at the end of curing, independent of the duration of curing, as shown by Fig. 7.5, may not hold true any more if the average permeability of an entire specimen is measured rather than that of the surface layer of a concrete cube, as has been done for the data shown in Fig. 7.5. Hydration continues in regions away from the surface even after termination of curing so that the strength at the end of curing no longer reflects the pore structure at the time of testing permeability. However, the approach taken in [7.7] and [7.8] is of particular significance for the correlation between strength and durability data. This will be dealt with in section 7.3.1.

7.2.2 CAPILLARY SUCTION

In Chapter 3 relations between different transport characteristics are dealt with. Rate of water absorption and water permeability may be correlated according to the relations given in section 3.2.3. Since from section 3.3.4 it follows that correlations also exist between gas permeability and water permeability, there should be a correlation between capillary suction and air permeability and as a consequence between capillary suction and standard compressive strength. However, these interrelations are so complex that no attempt has been made to express them in analytical terms.

Experimental data for rate of water absorption and standard compressive strength are given e.g. in [7.4–7.6]. In Fig. 7.6 the rate of water absorption a (g/m^2 s$^{0.5}$) is given as a function of standard compressive strength (MPa) for the data reported in [7.4]. The rate of absorption has been determined on cylinders 50 x 50 mm, which had been moist-cured

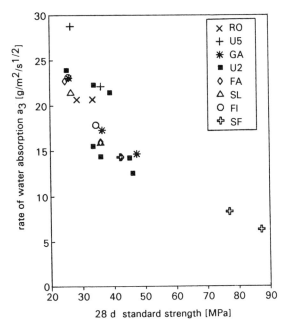

Fig. 7.6. Relation between rate of absorption of concrete and standard strength for concretes made of different types of cement [7.4].

for 7 days and preconditioned by oven-drying at 50°C for a period of 6 days and tested at an age of 35 days. The rate of water absorption was determined after a suction time of 3 h. According to Fig. 7.5 there exists a reasonably good correlation between rate of absorption and compressive strength on a linear scale, which is independent of the type of cement.

The corresponding data from [7.5] and [7.6] are given in Fig. 7.7. The concretes investigated correspond to those for which permeability data are presented in Figs 7.3 and 7.4. The absorption rate was determined after a suction time of 3 min on specimens that, after curing, had been air-dried at 65% RH and $T = 20$°C up to an age of 56 days. In Fig. 7.7 distinction is made between the rate of absorption of specimens that had been moist-cured for 1, 3 and 7 days, respectively. Similar to the permeability coefficients, the correlation between rate of absorption and average compressive strength is independent of the type of cement but strongly influenced by the duration of curing. As the duration of curing is increased the rate of water absorption decreases substantially.

The effect of curing on water absorption also follows from Fig. 7.8, which is taken from [7.9]. The increase in initial surface absorption (ISA), as defined in section 9.3, beyond the ISA value for water-cured concrete is given as a function of compressive strength for specimens that had been subjected to the following curing regimes:

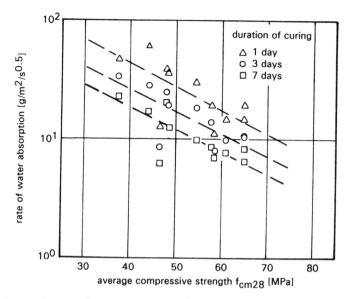

Fig. 7.7. Rate of absorption of concrete and compressive strength for concretes made of different types of cement and cured for 1, 3 or 7 days [7.6].

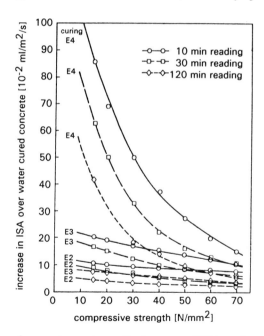

Fig. 7.8. Influence of concrete grade on the increase of ISA under various curing conditions with respect to water-cured concrete [7.9].

E2 6 days water, then air at 20°C, 55% RH
E3 3 days water, then air at 20°C, 55% RH
E4 Air at 20°C, 55% RH

As the duration of curing is increased the ISA values for a given strength grade decrease substantially. It does not follow from [7.9] whether unique or less diverging relations will be obtained if the specimens to determine compressive strength and those to determine the ISA values are cured identically.

The conclusions drawn from the experimental results on absorption rate are similar to those for the permeability data: unique correlations between strength and absorption rate can be expected only if the specimens are cured identically and tested at the same age and if the moisture states at the time of testing are controlled by suitable preconditioning.

7.2.3 DIFFUSION

In Chapter 3 of this report relations between the water vapour diffusion coefficient D_c and water permeability K_w as well as between gas diffusion coefficients D and gas permeability K_g are given. They have the general form:

$$K = \text{constant} \times D^b$$

Considering the relation between water permeability and compressive strength of the paste as expressed by eq. (7.3) and between gas permeability and strength as expressed by eq. (7.6), relations between diffusion coefficients and concrete strength should have the general form:

$$D = \text{constant} \times (1/f_{cp})^s \tag{7.9}$$

In investigations where diffusion coefficients for water vapour or other gases have been determined, generally concrete or paste strength have not been reported. Therefore, eq. (7.9) cannot be verified on the basis of experimental data. It is to be expected that the same conclusions that have been drawn for the correlation between permeability and strength are also valid for the correlation between diffusion coefficients and strength.

Equation (7.9) in its general form can be valid only in cases in which the pore structures of the specimens for diffusion testing as well as for strength testing are similar. In addition, the effects of a particular moisture state on diffusion coefficients can never be estimated on the basis of a standard strength value.

Nevertheless, some correlations have been established between diffusion-controlled characteristics such as drying shrinkage of concrete and compressive strength. Figure 7.9 taken from [7.10] gives a correlation

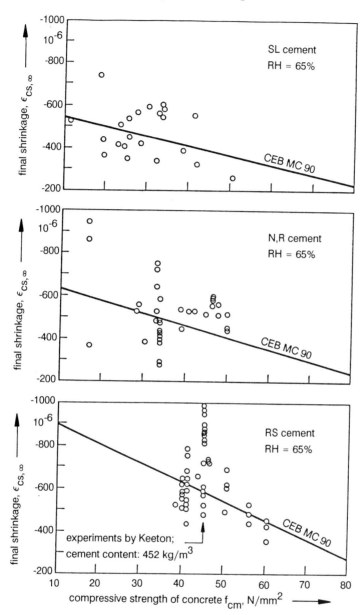

Fig. 7.9. Relations between the final shrinkage and the compressive strength of concretes made with slowly hardening (top), normal or rapid hardening (middle) and rapid-hardening high-strength cements (bottom), from [7.10].

between shrinkage and the respective strength of the concrete. As the compressive strength increases, shrinkage of concrete decreases. In this diagram the effects of member size, relative humidity and type of cement have been taken into account. Even then, this correlation exhibits substantial scatter, because other parameters such as water and cement content are of equal significance.

7.3 Compressive strength and durability

7.3.1 CARBONATION

In numerous experimental investigations relations between depth of carbonation and compressive strength have been investigated, e.g. [7.11–7.17]. In all cases there is a general trend that the depth of carbonation decreases with increasing compressive strength. The data from [7.5], [7.6] and [7.11], which are shown in Figs 7.10–7.13, are also representative for the results obtained in other investigations, which are not dealt with in detail in this chapter. In the study reported in [7.11] concretes made of four types of cement, water/cement ratios of 0.45, 0.60 and 0.80 and cured for 7 days have been exposed to carbonation either in the laboratory at 65% RH, 20°C, outdoors sheltered from rain or outdoors horizontally exposed to

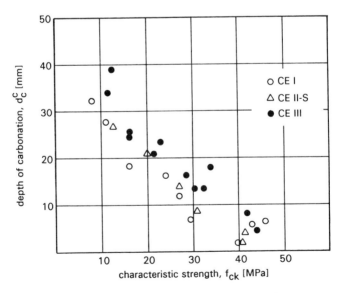

Fig. 7.10. Depth of carbonation after 16 years of storage in a constant environment of 20°C, 65% RH versus compressive strength at an age of 28 days, duration of curing 7 days: data from [7.11].

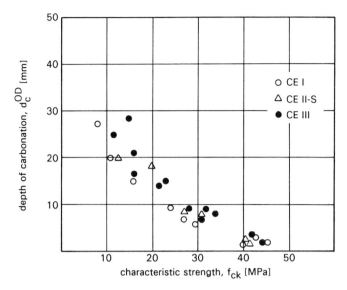

Fig. 7.11. Depth of carbonation after 16 years of storage outdoors, sheltered from rain versus compressive strength at an age of 28 days, duration of curing 7 days: data from [7.11].

rain over a period of 18 years. In Figs 7.10–7.12 the depth of carbonation observed after 18 years of exposure is given as a function of the character-istic compressive strength f_{ck} of concrete cubes, side length 200 mm, cured for 7 days and tested at an age of 28 days. Though the depth of carbon-ation decreases with increasing f_{ck} for all types of concrete, these relations depend on the type of cement, with concretes made of blastfurnace slag cements showing higher depths of carbonation for a given characteristic compressive strength.

The effect of curing on carbonation–strength relations has been studied, e.g. in [7.5] and [7.6]. In Fig. 7.13 the depth of carbonation after 1 year of exposure in the laboratory at 20°C and 65% RH is given as a function of the average compressive strength of cubes, side length 150 mm and cured for 7 days. For a given concrete grade the depth of carbonation decreases substantially as the duration of curing is increased.

In order to reduce or eliminate the effect of curing on the carbonation–strength relationship it was proposed in [7.7], [7.8] and [7.18] to relate depth of carbonation after a given exposure time to the concrete compressive strength at the end of curing, similar to the presentation of permeability data given in Fig. 7.5. This approach appears to be very logical, since carbonation is controlled by the pore structure of the near-surface regions of the concrete, i.e. the properties of the cover concrete. Under the assump-tion that concrete hydration stops soon after termination of curing,

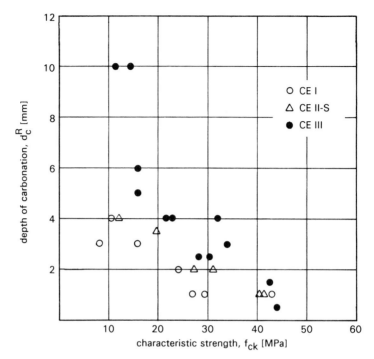

Fig. 7.12. Depth of carbonation after 16 years of storage outdoors horizontally exposed to rain versus compressive strength at an age of 28 days, duration of curing 7 days [7.11].

concrete compressive strength at the end of curing should reflect the pore structure of the cover concrete even at a higher age. In Fig. 7.14 depth of carbonation after 1.5 years is given as a function of compressive strength at the end of curing (f_{cmc}) on a double logarithmic scale, for the data from [7.18]. Figure 7.15 gives the same carbonation data versus compressive strength at an age of 28 days. These relations may be described by

$$d_c = d_{co} \cdot \left(\frac{f_{cm}}{f_{cmo}} \right)^b$$

(7.10)

where d_c =   depth of carbonation (mm)
      d_{co} =   coefficient (mm)
      f_{cmo} =   1 MPa
      b =   exponent

From a least-squares fit the coefficients d_{co} as well as the exponent b have been determined both for the d_c–f_{cm28} and for the d_c–f_{cmc} relation. They are given in Table 7.1.

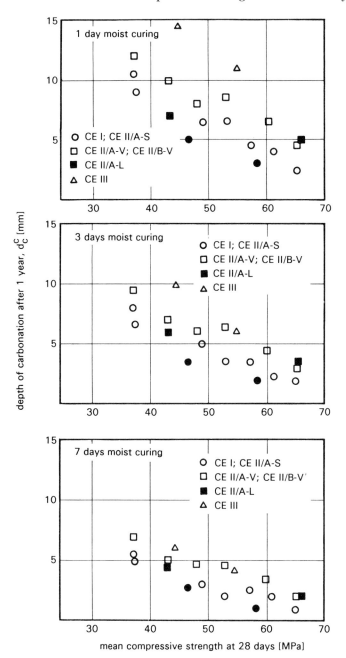

Fig. 7.13. Depth of carbonation after 1 year of exposure in a constant environment of 20°C, 65% RH versus concrete compressive strength f_{cm28} [7.6].

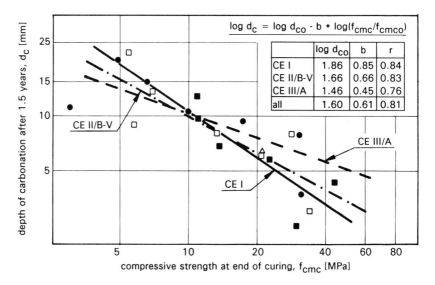

Fig. 7.14. Depth of carbonation after 1.5 years of storage in a constant environment of 20°C, 60% RH versus compressive strength at the end of curing, f_{cmc}: data from [7.7].

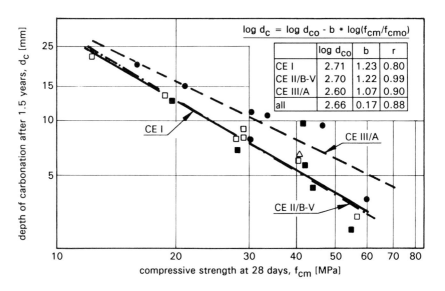

Fig. 7.15. Depth of carbonation after 1.5 years of storage in a constant environment of 20°C, 60% RH versus compressive strength at an age of 28 days, f_{cm28}: data from [7.7].

Table 7.1

Type of cement	$d_{co}(mm)$		b		Correlation coefficient	
	f_{cmc}	f_{cm28}	f_{cmc}	f_{cm28}	f_{cmc}	f_{cm28}
CE I	72.55	512	−0.84	−1.23	0.84	0.80
CE II / B–V	45.7	490	−0.66	−1.22	0.83	0.80
CE II / A	28.8	398	−0.45	−1.07	0.76	0.91
all types	39.8	457	−0.61	−1.17	0.81	0.88

From Table 7.1 it follows that the correlation coefficients for the relationships $d_c = f\,(f_{cmc})$ are generally lower than for the relations $d_c = f_c\,(f_{cm28})$. From this one might conclude that the compressive strength at the end of curing is not a substantially better parameter for the prediction of depth of carbonation than a standard strength f_{cm28}. However, the database for Figs 7.14 and 7.15 is rather limited.

Unfortunately, for most carbonation data found in the literature, the strength at the end of curing is not known. Therefore, for the carbonation data from [7.5] and [7.6] that are presented in Fig. 7.13, specimens for additional strength tests have been cast using the same constituent material that had been used for the specimens from which the carbonation data were obtained. The standard compressive strength at an age of 28 days was almost identical for both series of experiments, so it is justified to correlate the carbonation data from [7.5] and [7.6] with the strength data from the second series of experiments. In Fig. 7.16 the carbonation data from Fig. 7.13 are given as a function of the compressive strength at the end of curing, f_{cmc}. In Fig. 7.16 distinction is made between concretes made of Portland cements (CEI) and blended cements with a low slag content (CEII-S) on the one hand, and concretes made of blastfurnace slag cements (CEIII) with a slag content of approximately 65% on the other hand. There is a close correlation between depth of carbonation, d_c, and f_{cmc}, irrespective of the duration of curing, resulting in correlation coefficients for the concretes made of Portland cements or blended cements with a low slag content of 0.90 and for the concretes made of blastfurnace slag cements of 0.91. However, for the slag cements, higher rates of carbonation are to be expected for a given compressive strength at the end of curing. The curves given in Fig. 7.16 follow eq. (7.10) with:

- for the Portland cements CEI, CEII-S:
  $b = -0.95$ and $d_{co} = 88.2$ (mm)
- for the slag cements CEIII:
  $b = -0.91$ and $d_{co} = 129.6$ (mm)

From this evaluation we may conclude that for a given type of cement

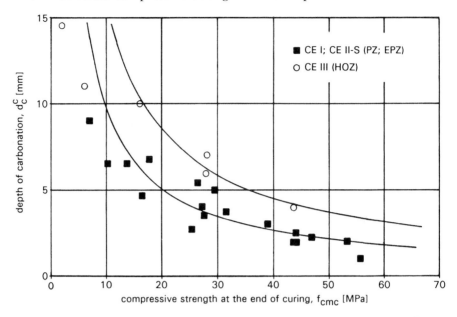

Fig. 7.16. Depth of carbonation after 1 year of exposure in a constant environment of 20°C, 65% RH versus concrete compressive strength at the end of curing (carbonation data from [7.5, 7.6]).

the effect of duration of curing on depth of carbonation–strength relationships will be eliminated or at least reduced if concrete compressive strength at the end of curing is used as a parameter.

7.3.2 FROST RESISTANCE

In the literature, as well as in national and international standards, frost resistance of concrete is related to strength only in so far as a certain minimum strength class of concrete, generally in the range of 25 MPa, is required for concrete structures exposed to severe freezing and thawing. In addition to the particular exposure conditions the governing parameters for the frost resistance of concrete are the amount of freezable water and, in cases where a high degree of water saturation is to be expected, the existence of an artificial air void system in the hydrated cement paste. The amount of freezable water depends on the capillary porosity, i.e. the water/cement ratio and the degree of hydration, as well as on the rate at which water can be absorbed by the concrete. Consequently continuity and discontinuity of the pore system may be decisive [7.2, 7.19].

Concrete compressive strength may reflect the effects of water/cement ratio, degree of hydration and capillary continuity. Thus, an increase of frost resistance with increasing concrete strength is to be expected to some extent, particularly for non-air-entrained concretes. However, the positive effect of air entrainment never reflects itself in the standard compressive strength; in fact the compressive strength decreases with increasing air content.

However, concrete compressive strength is a good and generally accepted indicator in determining the age of young concrete at which it may be exposed to freezing. Powers proposed a minimum strength of about 3.5 MPa (500 psi) [7.19, 7.20]. In [7.21] a minimum maturity that has to be reached prior to exposure to frost has been postulated. It depends on the water/cement ratio. Similar to Powers' approach, it is based on the assumption that prior to exposure to freezing, the capillary pores of the cement paste that were originally filled with mixing water have to be sufficiently emptied by self-desiccation so that the degree of saturation of the concrete is below a critical value, provided no additional water could penetrate the concrete.

The effect of concrete compressive strength is of particular relevance when estimating the frost resistance of high-strength concrete. In this context the question arises whether air entrainment is needed for such concretes because of their low capillary porosity and their very low rates of water absorption, which in turn lead to very low amounts of freezable water: e.g. [7.22]. In Fig. 7.17 from [7.23] the relative frost resistance of non-air-entrained concretes is given as a function of concrete compressive strength. Distinction is made between concretes that were allowed to dry

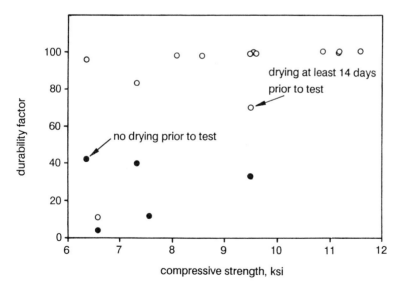

Fig. 7.17. Relative frost resistance and concrete compressive strength, from [7.22]

prior to exposure to freezing and thawing and concretes that were kept moist up to the time of testing. From this it follows that the effect of the moisture state of the concretes at the beginning of exposure to freezing and thawing by far overrides the influence of concrete compressive strength.

7.3.3 ABRASION RESISTANCE

(a) Mechanisms of abrasion

Abrasion or wear of concrete surfaces may be found in concrete pavements or industrial floors exposed to traffic or other moving loads. Also, the interior surfaces of concrete bins or silos may suffer abrasion after exposure to moving or dropping bulk material during charge or discharge. In hydraulic structures, sand and gravel may abrade the surfaces of the structure in addition to cavitation erosion caused by the flow of water. Thus, in contrast to other durability-related concrete properties, abrasion resistance may be considered a mechanical property.

As shown in [7.24] particles acting on a concrete surface by sliding, rolling or impact generate a multiaxial state of stress, which may eventually lead to local failure. In the vicinity of a particle that imposes compressive and shear stresses on a concrete surface, biaxial tension–compression stresses may cause tensile cracks in the concrete surface layer. Depending on the size of the particle and its motion relative to the concrete surface, as well as on the properties of the aggregates, fracture and thus abrasion is initiated either in the hydrated cement paste matrix, in the aggregates or in the paste–aggregate interface. Considering such a process it is obvious that close correlations should exist between abrasion resistance and other mechanical properties of the concrete, such as compressive or tensile strength, provided they are representative of the properties of the near-surface regions exposed to abrasion.

One of the many problems in defining concrete resistance to abrasion is the choice of a suitable test method. A thorough survey of such test methods can be found in [7.24]. The test methods may simulate a more or less uniform grinding action e.g. ASTM-C-779, Method A [7.25], the rolling action of wheels in medium traffic e.g. the dressing wheel according to ASTM-C-779 Method B [7.25] or of heavy traffic e.g. ASTM C-779, Method C [7.25] or ASTM-C-944 [7.26]. When applying shot blast or sand blast methods as described e.g. in ASTM-C-418 [7.27] primarily the hardness of the mortar matrix is tested. Also Vickers hardness, Brinell hardness and microhardness of the concrete surface may be used as an indicator of abrasion resistance [7.24]. The simultaneous effects of rolling particles and impact are simulated by methods such as the Talbot–Jones rattler test [7.29] and to some extent also by a rolling wheel type machine described in [7.30]. Special methods have been developed to simulate moving traffic or the effect of studded tires more closely e.g. [7.31, 7.32]. Methods to test the abrasion resistance of aggregates are given e.g. in ASTM-C-131 [7.33].

(b) Parameters influencing abrasion resistance

As long ago as 1919 and 1921, Duff A. Abrams published the results of a remarkable and extensive experimental investigation, which included abrasion tests on about 10 000 concrete specimens [7.28, 7.29]. The main parameters that were varied and which are of significance in this context were the water/cement ratio, duration of curing as well as type and grading of aggregates. For abrasion testing, the Talbot–Jones rattler was employed, which consists essentially of a vertical circular cast iron head. The concrete test specimens are attached to the inside of a heavy cylindrical band so that they form a 10-sided polygon. After the cylindrical band is set in the machine head the test chamber is filled with cast iron balls and closed with a steel plate. Upon rotation of the cast iron head around its horizontal axis the specimens are subjected to the tumbling action and impact of the cast iron balls. The experiments showed the strong influence on abrasion resistance of water/cement ratio, the duration of curing, as well as of the grading of aggregates. Wear of the exposed concrete surfaces decreased with decreasing water/cement ratio, increasing duration of curing and increasing fineness modulus of the aggregates i.e. a reduction of the sand content of the concrete. Since all of the aggregates investigated 'were of high grade' no pronounced difference in the abrasion resistance of concretes made of different types of aggregates could be found. Abrams also indicated that abrasion was more pronounced if the specimens were 'tested before they were thoroughly dry'. In general, variations of these parameters reflected themselves both in abrasion resistance and concrete compressive strength. This can be seen from Fig. 7.18, taken from [7.29], where concrete compressive strength prior to abrasion testing at an age of 120 days and wear after 1800 revolutions is given as a function of the duration of curing.

The results of extensive studies on the effect of type of aggregate on abrasion resistance of concrete are given in [7.34]. In these experiments three different methods for abrasion testing of the concrete were employed, and also the abrasion resistance of the aggregates themselves was determined. A total of 15 different types of natural aggregates and mixtures thereof were used to produce concrete mixes with water/cement ratios ranging from 0.32 to 0.70. All specimens were cured in an identical manner. An evaluation of the test data shows that for a given water/cement ratio, abrasion of concretes made of hard aggregates such as quartz or basalt is almost identical to the abrasion of concretes made of soft limestone aggregates if shot blasting is used as a test method. When

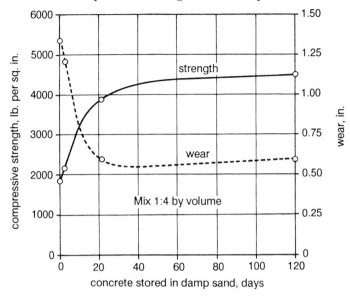

Fig. 7.18. Effect of duration of curing on concrete abrasion resistance and on concrete compressive strength, from [7.28]

applying a dressing wheel or a steel ball method similar to ASTM-C-779, Methods B and C, abrasion resistance of the concretes increased markedly with increasing abrasion resistance of the aggregates. However, the effect of aggregates diminishes with increasing concrete compressive strength, i.e. decreasing water/cement ratio. Also refer to Figs 7.21 to 7.23 in the following section.

According to the results given in [7.35], air entrainment influences abrasion resistance only in so far as it affects the compressive strength of concrete.

The influence of curing on abrasion resistance has been studied in several investigations e.g. [7.28–7.30, 7.36]. Figure 7.19 from [7.30] shows the influence of duration of curing in water and of the water/cement ratio on abrasion. From this it follows that curing becomes more critical as the water/cement ratio increases. A substantial improvement of abrasion resistance was also obtained if effective curing compounds were applied to the concrete surface. Similar results are reported in [7.36]; however, the improvement in abrasion resistance due to the use of curing compounds compared to the abrasion resistance of specimens without curing was less pronounced than reported in [7.30].

(c) Abrasion resistance and concrete strength

As indicated in section 7.3.3(a) the action of abrasion generates a multiaxial stress state and biaxial tension compression in the exposed surface, so

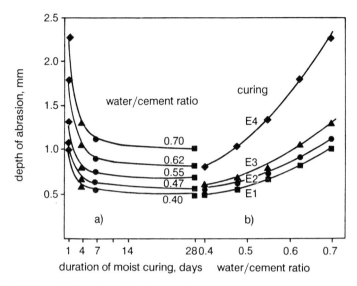

Fig. 7.19. Effect of duration of curing and water/cement ratio on abrasion resistance of concrete, from [7.29]

that close correlations between abrasion resistance and other strength criteria are to be expected. In none of the investigations on abrasion resistance was a strength criterion other than the compressive strength at the time of abrasion testing determined, so that interpretations of abrasion are limited to this mechanical property.

Already Abrams underlined the dominant influence of concrete compressive strength on abrasion resistance [7.28, 7.29]. From his studies he concludes that the following relation exists between compressive strength S and wear W:

$$S = A/Wn \qquad (7.11)$$

where A and n are constants that depend on the quality of the concrete and on the test method. In Fig. 7.20, taken from [7.29], compressive strength is given as a function of wear of concretes made of two types of aggregate, four curing conditions and six water/cement ratios. Equation (7.11) with $A = 2230$ and $n = 1.07$ describes the data given in Fig. 7.20 very well. Abrams states in this context: 'This relation appears to hold whether variations in strength are due to quantity of cement, size and grading of the aggregate, quantity of mixing water, age of concrete, curing condition of concrete, or other factors which affect strength. This relation appears to be generally true also for aggregates of different types.'

The latter statement is only partially supported by the data reported in

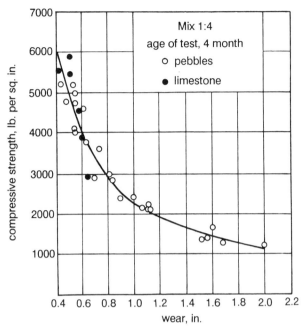

Fig. 7.20. Relation between wear and compressive strength of concrete, from [7.28]

[7.34]. In Figs 7.21–7.23, abrasion in g of abraded material is given for Test Series III of this investigation as a function of concrete compressive strength of concrete cylinders 76 x 152 mm, which were moist-cured up to the time of testing. The aggregates used were identified as follows: hard aggregate, soft aggregate, hard sand and soft gravel, hard gravel and soft sand. Figure 7.21 gives this relation for the results obtained with shot blasting as an abrasion test method. Figures 7.22 and 7.23 give the corresponding data for testing with steel balls and with a dressing wheel, respectively. A close correlation between abrasion and compressive strength only exists for the tests with shot blasting. As pointed out in section 7.3.3(a), with such a method the resistance of the mortar matrix at the surface of the specimen exposed to abrasion is tested. Consequently the abrasion resistance of the aggregates influences the abrasion resistance of the concrete only in so far as it affects its compressive strength. For the abrasion data obtained with the other two methods, the abrasion resistance of the aggregates influences the abrasion properties of the concrete much more than the compressive strength, so that, for a given compressive strength of the concrete, abrasion increases markedly as the abrasion resistance of the coarse aggregates is reduced. The influence of aggregates, however, on the abrasion–strength relation becomes almost negligible for a concrete strength exceeding about 55 MPa. The high

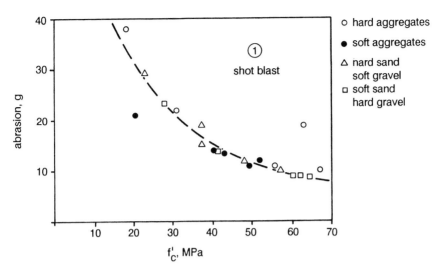

Fig. 7.21. Relation between abrasion and compressive strength of concrete: data from [7.33]. Test method: shot blasting.

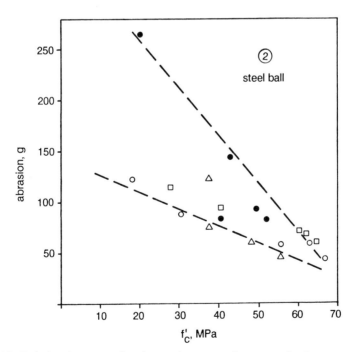

Fig. 7.22. Relation between abrasion and compressive strength of concrete: data from [7.33]. Test method: steel balls.

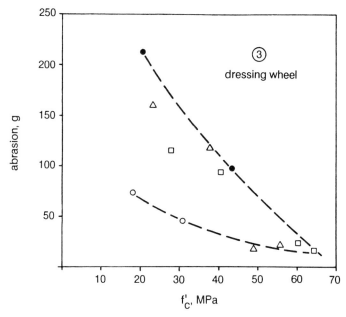

Fig. 7.23. Relation between abrasion and compressive strength of concrete: data from [7.33]. Test method: dressing wheel.

abrasion resistance of high strength concrete has also been pointed out in [7.24] and [7.31].

A unique abrasion–compressive strength relation when employing shot blasting for abrasion testing is also reported in [7.35] for air-entrained concretes.

The influence of curing on abrasion–strength relationships is more controversial. According to the data obtained by Abrams and as shown in Figs 7.18 and 7.19 curing influences abrasion resistance and compressive strength in a rather similar way. In Fig. 7.24, which is taken from [7.30], abrasion as obtained with a rolling wheel type abrasion machine is given as a function of concrete compressive strength for concretes made of different water/cement ratios and cured for different durations. Also, the maximum aggregate size, the workability of the fresh concrete as well as type of cement and additions have been varied. The correlation appears to be reasonable, with a correlation coefficient $r = 0.85$. In particular, the data obtained from specimens where the duration of curing was the only variable appear to follow a fairly unique relation. The conclusion given in [7.30] that 'slabs that have undergone different curing, but have equal strength, could possess very different abrasion resistances' appears to be slightly exaggerated. For the data shown in Fig. 7.24 the deviations from a unique relation between abrasion and strength occur primarily for the

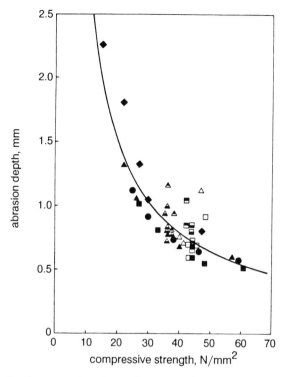

Fig. 7.24. Relation between abrasion and compressive strength of concrete, from [7.29].

abrasion of concretes with different workability, which resulted in different bleeding properties and consequently markedly different surface characteristics. It should be pointed out that for the same concretes the relation between abrasion and ISA is given in Fig. 5.17 of Chapter 5 of this state-of-the-art report. A considerably higher correlation coefficient, $r = 0.95$, was obtained.

7.4 Conclusions

From this brief literature survey on strength–permeability–durability relations the following conclusions may be drawn:

1. For hydrated cement pastes as well as for concrete permeability–compressive strength correlations should exist that are little affected by parameters such as age, duration of curing or type of cement, provided strength and permeability are determined on specimens with identical

pore structures and moisture states. Under such conditions gas or water permeability may be estimated from compressive strength data. This should also hold true for relations between rate of absorption and compressive strength.

2. The correlation between permeability and a standard compressive strength, which was determined at a given age of concrete, depends on the duration of curing: as the duration of curing is increased, gas or water permeability as well as rate of absorption decrease substantially.

3. Rate of carbonation–strength relationships depend both on duration of curing and type of cement. For a given strength grade the depth of carbonation at a given concrete age decreases as the duration of curing increases. For a given strength grade and a given duration of curing, the depth of carbonation at a given age is higher for concretes made of blastfurnace slag cements with a high slag content than for concretes made of Portland cements.

4. The effect of curing on permeability–strength relationships as well as on carbonation–strength relationships may be reduced or eliminated if a standard compressive strength at a given age is replaced by the compressive strength at the end of curing.

5. Compressive strength is not a reliable indicator of the resistance of concrete to freezing and thawing as long as the compressive strength exceeds values of approximately 25 MPa.

   The lack of a unique relation between frost resistance and compressive strength is demonstrated by the positive effects of entrained air: as the air content increases the frost resistance increases, whereas concrete compressive strength decreases. However, for high-strength concrete a high resistance to freezing and thawing may be expected because of the low amounts of freezable water and because of the high resistance of such concretes to the penetration of water.

6. Correlations between abrasion and compressive strength are markedly influenced by the type of abrasive action or the test method employed. A fairly unique correlation between abrasion resistance and concrete strength exists if the concrete surface is subjected to a more uniform abrasion by small particles such as shot or sand blasting. However, this relation is strongly influenced by the abrasion resistance of the aggregates if the abrasive action is more local, as is the case in surfaces exposed to moving vehicles.

7. An abrasion–compressive strength relation depicts the effects of technological parameters such as water/cement ratio, grading of aggregates and to some extent duration of curing on abrasion resistance well. However, it does not reflect variations of parameters that influence only the properties of the surface layer of a concrete member, such as effects of bleeding, or special surface treatments.

7.5 References

7.1. Powers, T.C. and Brownyard, T.L. (1948) Studies of the physical properties of hardened Portland cement paste. *Portland Cement Association, Research Bulletin*, No. 22.

7.2. Powers, T.C. (1958) The physical structure and engineering properties of concrete. *Portland Cement Association, Research Bulletin*, No. 90.

7.3. Powers, T.C., Copeland, L.E., Hayes, J.C. and Mann, H.M. (1954) Permeability of Portland cement paste. *Journal of the American Concrete Institute, Proceedings*, Vol. 26, No. 23, pp. 285–298.

7.4. Torrent, R.J. and Jornet, A. (1991) The quality of the 'covercrete' of low-, medium- and high-strength concretes, Second CANMET/ACI International Conference, *Durability of Concrete*, Montreal, Canada, August.

7.5. Schönlin, K.F. (1989) Permeabilität als Kennwert der Dauerhaftigkeit von Beton. *Schriftenreihe des Instituts für Massivbau und Baustofftechnologie, Universität Karlsruhe*, Heft 8.

7.6. Hilsdorf, H.K., Schönlin, K.F. and Burieke, F. (1991) *Dauerhaftigkeit von Betonen*, Institut für Massivbau und Baustofftechnologie, Universität Karlsruhe.

7.7. Parrott, L. (1990) Carbonation, corrosion and standardization, in *Protection of Concrete*, Proceedings of the International Conference, University of Dundee, Eds. R.K. Dhir and J.W. Green, E. & F.N. Spon, London, pp. 1009 23.

7.8. Parrott, L. and Hong, Ch.Z. (1990) Some factors influencing air permeation measurements in cover concrete. *Materials and Structures*, Vol. 24, No. 144, November, pp. 403–408.

7.9. Dhir, R.K., Hewlett, P.C. and Chan, Y.N. (1987) Near-surface characteristics of concrete: assessment and development of in situ test methods. *Magazine of Concrete Research*, Vol. 39, No. 141, December, pp. 183–195.

7.10. Comité Euro-International du Beton – CEB (1990) *Evaluation of the time dependent behaviour of concrete*, Bulletin D'Information No. 199, Lausanne.

7.11. Wierig, H.J. (1984) Longtime studies on the carbonation of concrete under normal outdoor exposure, in *Proceedings of the RILEM Seminar on the Durability of Concrete Structures under Normal Outdoor Exposure*, Institut für Baustoffkunde und Materialprüfung, Universität Hannover, pp. 239–249.

7.12. Wierig, H.J. and Scholz, E. (1989) *Untersuchungen über den Einfluß von Flugaschezusätzen auf das Carbonatisierungsverhalten von Beton, 1. Ergänzung*, Schlußbericht IRB Verlag, Stuttgart.

7.13. Report of RILEM Technical Committee 16-C (1978) Carbonation. *Matériaux et Constructions*, Vol. 11, No. 62, pp. 142–146.

7.14. Smolcyk, H.G. (1968) Discussion to the problem 'Carbonation of concrete', in *Proc. Fifth International Symposium on the Chemistry of Cement*, Tokyo, Written discussion to principal paper No. III-3.

7.15. *Schriftenreihe des Deutschen Ausschuß für Stahlbeton* (1965) H. 170, Verlag Wilhelm Ernst & Sohn, Berlin, mit Beiträgen über 'Beobachtung an alten Stahlbetonbauteilen hinsichtlich Carbonatisierung des Betons und Rostbildung an der Bewehrung', über 'Untersuchungen über das Fortschreiten der Carbonatisierung an Betonbauwerken' und über 'Tiefe der carbonatisierten Schicht alter Betonproben, Untersuchungen an Betonproben'.

7.16. Nischer, P. (1984) Einfluß der Betongüte auf die Karbonatisierung. *Zement und Beton*, Vol. 29, No. 1, pp. 11–15.

7.17. Fattuhi, N.J. (1986) Carbonation of concrete as affected by mix constituents and initial water curing period. *Matériaux et Constructions*, Vol. 19, No. 110, pp. 131–136.

7.18. Parrott, L.J. (1990) Water absorption in cover concrete. *Materials and Structures*, Vol. 25, No. 149, June, pp. 284–292.

7.19. ACI Committee 201 Durability (1991) Guide to Durable Concrete. *ACI Materials Journal*, September/October.

7.20. Powers, T.C. (1962) Prevention of frost damage of green concrete. *Bulletin RILEM*, No. 14, Paris.

7.21. Comité Euro-International du Beton – CEB (1989) *Durable concrete structures*, CEB design guide, Bulletin D'Information No. 182, Lausanne.

7.22. FIP–CEB Working Group on high strength concrete (1990) *High Strength Concrete – State of the art report*, CEB Bulletin D'Information No. 197, Lausanne.

7.23. Fiorato, Th.E. (1989) PCA research on high strength concrete. *Concrete International*, April, pp. 44–50.

7.24. Kunterding, H. (1991) Beanspruchung der Oberfläche von Stahlbetonsilos durch Schüttgüter, *Schriftenreihe des Instituts für Massivbau und Baustofftechnologie*, Universität Karlsruhe, Heft 12.

7.25. ASTM-C-779-82 Standard Test Method of Abrasion Resistance of Horizontal Concrete Surfaces.

7.26. ASTM-C-944-80 Standard Method of Abrasion Resistance of Concrete or Mortar Surfaces by the Rotating-Cutter Method.

7.27. ASTM-C418-81 Standard Test Method of Abrasion Resistance of Concrete by Sandblasting.

7.28. Abrams, D.A. (1919) Effect of curing condition on the wear and strength of concrete, in *Proceedings, American Railway Engineering Association*, Vol. 2.

7.29. Abrams, D.A. (1921) Wear tests of concrete. *ASTM Proceedings*, Vol. 12, pp. 1013–1034.

7.30. Dhir, R.K., Hewlett, P.C. and Chan, Y.N. (1991) Near-surface characteristics of concrete: abrasion resistance. *Materials and Structures*, Vol. 24, pp. 122–128.

7.31. Gjørv, O.E., Baerland, T. and Rønning, H.R. (1987) High strength concrete for highway pavements and bridge decks, in *Utilization of High Strength Concrete*, Proceedings, Stavanger, Ed. Holand, I., Trondheim, pp. 111–121.

7.32. Springenschmid, R. and Sommer, H. (1971) Untersuchungen über die Verschleißfestigkeit von Straßenbeton bei Spikesreifen-Verkehr. *Straße und Autobahn*, Heft 4, pp. 136–141.

7.33. ASTM-C-131-81 Standard Test Method of Resistance to Degradation of Small-Size Coarse Aggregate by Abrasion and Impact in the Los Angeles Machine.

7.34. Smith, F.L. (1958) Effect of aggregate quality on resistance of concrete to abrasion, in *Cement and Concrete*, ASTM STP 205, pp. 91–105.

7.35. Witte, L.P. and Backstrom, J.E. (1951) Some properties affecting the abrasion resistance of air-entrained concrete. *ASTM Proceedings*, Vol. 51, pp. 1141–1154.

7.36. Senbetta, E. and Malchow, G. (1987) Studies on control of durability of concrete through proper curing, in *Concrete Durability – Katherine and Bryant Mather International Conference*, ACI SP-100, pp. 73-87, Detroit, Michigan.

8 Microstructure and transport properties of concrete

Edward J. Garboczi

8.1 Introduction

Concrete is a composite material whose microstructure is random over a wide range of length scales. At the largest length scale, concrete can be considered to be a mortar–rock composite, where the randomness in the structure is on the order of centimetres, the size of a typical coarse aggregate. Mortar itself can be considered to be a cement paste–sand composite, with random structure on the order of millimetres. Cement paste can also be considered to be a random composite material, made up of unreacted cement, CSH, CH, capillary pores, and other chemical phases. The randomness in the cement paste microstructure is on the order of micrometres. Finally, CSH is itself a complex material, with random structure, as seen by neutron scattering [8.1], on the order of nanometres. This range of random structure, from nanometres (CSH) to centimetres (concrete) covers seven orders of magnitude in size! It is a large and difficult task to try to relate microstructure and properties theoretically for concrete. However, there are some simple, basic ideas that do provide a framework for this task, with the main difficulty being carrying these ideas through to specific application.

This chapter attempts to outline the general principles that must be considered in trying to understand microstructure transport property relationships in concrete, or indeed any other random porous material. Specific applications to cement paste, mortar, and concrete will be considered. An earlier review [8.2], which mentions some of the ideas discussed in this chapter, is also a helpful reference for some of the earlier transport property data and their interpretation in terms of pore structure.

8.2 Basic concepts

Typical transport processes important in concrete are transport of water under a hydrostatic pressure head, transport of water by capillary suction,

Performance Criteria for Concrete Durability, edited by J. Kropp and H.K. Hilsdorf. Published in 1995 by E & FN Spon, London. ISBN 0 419 19880 6.

the diffusion of ions under an applied concentration gradient, and the transport of ions by moving water. For these kinds of properties, two very simple ideas control how the spatial geometry of the microstructure affects transport properties. These ideas can be expressed in terms of **tube theory**: (1) large-diameter tubes have higher transport rates than small-diameter tubes, and (2) tubes that are blocked have zero transport rates. These ideas, phrased more rigorously as **pore size** and **connectivity**, provide the theoretical framework necessary for describing how the transport properties of concrete depend on pore structure. The idea of connectivity will be discussed first.

8.3 Description of connectivity by percolation theory

The ideas of percolation theory are very helpful in understanding the important features of random structures. The main concept of percolation theory is the idea of **connectivity**. Picture some sort of structure being built up inside a box by randomly attaching small pieces to a pre-existing central core. Percolation theory attempts to answer the question: at what point does the structure span the box? An alternative form of this question, for a random structure that already spans the box, is: if pieces of the structure are removed at random, when will it fall apart? The percolation **threshold** is defined by the value of some parameter, say volume fraction, right at the point where the structure either achieves or loses continuity across the box.

Figure 8.1 shows a simple two-dimensional example of these ideas. Circular blobs of 'paint' are thrown down at random on a clean sheet of paper, and the area fraction covered by the paint is monitored until the paint blobs form a continuous structure. It is found numerically that the paint blobs will become continuous when they cover approximately 68% of the paper [8.3]. In Fig. 8.1 the circles have an area fraction of 72% and so form a connected path. This is an example of the percolation of a structure that is being randomly built up. If we now think of the paper as a conducting sheet, and the paint blobs as circular holes that are randomly punched out, then the sheet will lose connectivity and its ability to carry an electrical current at an area fraction of 32%. This is an example of the percolation of a structure that is being randomly torn down. Concrete exhibits both kinds of percolation processes, as discussed further below.

8.4 Pore size effects on transport properties

As was stated in the introduction, it is obvious that the wider the tube carrying the flow, the greater the flow will be. One addition to this obvious

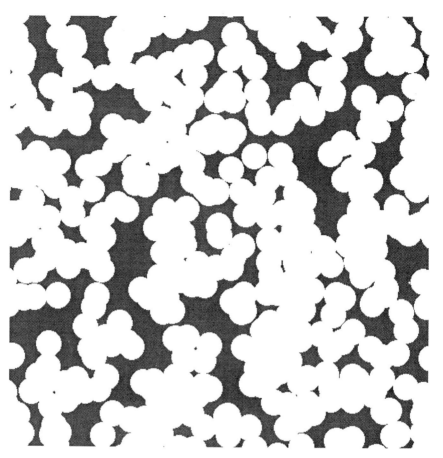

Fig. 8.1. Showing a unit cell of a system of randomly centred monosize circles, where 72% of the area is covered by the circles. The circles form a continuous phase.

notion, however, is that the tube or pore size will affect different transport properties in different ways. For example, consider an L x L x L cube of porous material, where the pore space is in the shape of a single tube of radius R, as in Fig. 8.2(a). The transport properties of such a simple material are easy to compute. The porosity is $\phi = \pi R^2/L^2$, the permeability is $k = \pi R^4/(8L^2)$ [8.4], and the ionic diffusivity, if the pore is filled with a fluid in which the ions of interest have an intrinsic diffusivity D_o, is $D = D_o \pi R^2/L^2$. Figure 8.2(b) shows an L x L x L cube of another material whose pores are in the shape of N ($N = 4$ in the figure) parallel tubes of radius $r = R/\sqrt{N}$. The porosity is the same, since in this case $\phi = N(\pi r^2/L^2) = \pi R^2/L^2$, but now there are more pores and they are smaller. The diffusivity will be unchanged as well, since $D = D_o N(\pi r^2/L^2) =$

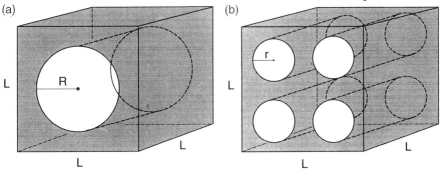

Fig. 8.2. Showing two simple models of a porous material, where the pores are straight, parallel tubes of uniform width passing through the solid material: (a) one tube of radius R, (b) $N = 9$ tubes of radius $r = R/\sqrt{N} = R/3$.

$D_o \pi R^2/L^2$. However, the permeability is now different and smaller, since $k = N[\pi r^4/(8L^2)] = (1/N)[\pi R^4/(8L^2)]$. This argument will be valid, in general, even for random pore structures, so that it is important to remember that fluid permeability and ionic diffusivity depend on pore size in different ways. They will depend on pore connectivity in a similar manner, however.

8.5 Cement paste

We now specifically consider the microstructure–transport property relationships in cement-based materials. It is important to begin with cement paste, as this is the matrix material that, along with sand and rock inclusions, forms the concrete composite. It is difficult to understand the behaviour of a composite without first understanding each constituent separately.

For simplicity, and because it still correctly captures the main features of the pore structure, cement paste can be thought of as consisting of four phases: (1) unreacted cement, (2) surface products like CSH, (3) pore products like CH, and (4) capillary pore space. Surface products grow out from the unreacted cement particles and contain continuous gel pores, while pore products are generally crystalline and fully dense, with no connected pores. The capillary pores are the left-over space between solid phases.

8.5.1 PORE CONNECTIVITY

Cement paste is itself a random porous material, with more than one distinct solid phase. Percolation ideas can be applied to each particular material phase, and the 'box' taken to be the macroscopic sample. The connectivity of the different phases, and particularly the pore phases, changes with time.

Immediately after mixing, the solid phases are discontinuous, and so the freshly mixed paste is a viscous liquid. The solid phase is then built up through random growth of reaction products, and at some point becomes continuous across the sample, mainly due to the formation of the CSH surface products [8.5]. This percolation threshold is then a rigorous theoretical definition of the set point.

A percolation threshold that is more important for transport processes is the point at which the capillary pore space loses continuity. Such a percolation threshold can exist, because as hydration products are formed, pieces of the capillary pore space will be trapped and cut off from the main pore network, thus reducing the fraction of the pores that form a connected pathway for transport. As this process continues, the capillary pore space can lose all long-range connectivity, so that 'fast' transport of water or ions through the relatively large capillary pore system would end, and 'slow' transport would then be regulated by the smaller CSH gel pores.

Results from recent computer simulation work provide evidence for such a percolation threshold. Figure 8.3(a) shows the 'fraction connected' of the capillary pore space vs degree of hydration for several water/cement ratios, as computed by a computer simulation model of cement paste microstructure [8.6]. The quantity 'fraction connected' is defined as the volume fraction of capillary pores that make up a connected path through

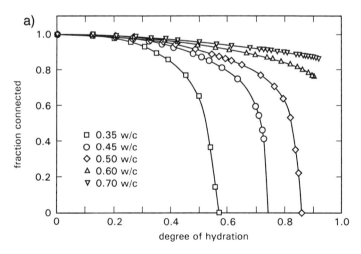

Fig. 8.3. Showing the fraction of the capillary pore space that is part of a perco-lated (continuous) cluster, for several different water/cement ratio cement pastes as a function of: (a) degree of hydration, and (b) capillary porosity.

the sample, divided by the total volume fraction of capillary pores. Immediately after mixing, the cement particles are totally isolated, assuming adequate dispersion, and so the connected fraction of the capillary pore space is 1. As hydration occurs the connected fraction decreases gradually. If continuity is lost at some critical degree of hydration, the 'fraction connected' will go to zero. Such a percolation threshold can be seen in all the water/cement ratios plotted, except for 0.6 and 0.7. We have found in the model that water/cement ratios of 0.6 and above always have a continuous (or percolated) capillary pore system. This prediction is in good agreement with experiment [8.7]. It is clearly seen in Fig. 8.3(a) that as the water/cement ratio decreases below 0.6, less and less hydration is required to close off the capillary pore system.

In order to unify the previous results, we have replotted all the data from Fig. 8.3(a) in Fig. 8.3(b) against capillary porosity. All the connectivity data now fall on one curve, and it is clearly seen that there is a common percolation threshold at a critical value of capillary porosity of about 0.18. Even the 0.6 and 0.7 water/cement ratio data fall on this curve, and now it is clear why these pastes always have an open capillary pore space: there is not enough cement present originally to be able to bring the capillary porosity down to the critical value, even after full hydration. The capillary pore space percolation threshold for cement paste will have some sensitivity to cement particle size distribution and degree of dispersion, so that the critical value of capillary porosity for percolation should be considered to be about $18 \pm 5\%$.

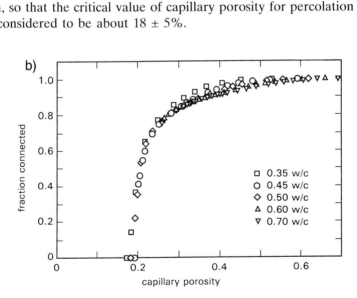

Fig. 8.3 (b).

8.5.2 PORE SIZE

During hydration, the capillary pore size, as well as the overall capillary porosity, decreases due to consumption of water during hydration, producing hydration products that fill in the capillary pore space. The size of the gel pores is fixed by the structure of the CSH, however, and so they remain constant in size. As hydration continues, the size of the capillary pores decreases down towards the roughly 10 nm gel pore size [8.2]. Therefore the importance of the pure capillary pore transport paths decreases with time due both to decreasing capillary pore size as well as to decreasing connectivity.

8.5.3 DIFFUSIVITY OF CEMENT PASTE

Building on the percolation and pore size results given above, the dependence of the diffusivity and permeability of cement paste on pore structure can now be qualitatively outlined. Early in hydration the capillary pore space is fully percolated. These pores are much larger than the CSH gel pores (which are themselves also fully connected fairly early in hydration [8.6]) and so dominate the transport. As the capillary porosity decreases, the capillary pores become smaller and only partially connected, so for porosities near but above the percolation threshold, pure capillary pore paths have only slightly more influence on flow than the hybrid paths that are made up of isolated capillary pockets linked by CSH gel pores. The capillary pores are still somewhat bigger than the gel pores, but their connectivity is becoming small. Below the critical capillary porosity, all flow must now go through CSH gel pores, but flow will be dominated by paths that contain some isolated capillary pore regions, and are not just made up of pure CSH gel pores. If this were not true, then after a certain point, the diffusivity would begin going up with increasing hydration, since more CSH and thus more gel pores were being formed. This is not the case in cement paste [8.8]. The same microstructure model as was used to predict the connectivity results shown in Fig. 8.3 can also be used to compute the diffusivity of cement paste by solving Laplace's equation in the simulated pore space with a finite difference method [8.8]. Computational results confirm the above microstructural picture, and compare reasonably well with experimental measurements [8.8, 8.9].

8.6 Microstructure of mortar and concrete

We now consider the effect of mortar and concrete microstructure on transport properties like ionic diffusivity and fluid permeability. For reasons of scale, mortar or concrete is considered as a composite, with the fine and coarse aggregate being the inclusions and cement paste considered as a uniform matrix. We average out the cement paste micrometre length scale to avoid having to deal with microstructure on too wide a range of length scales at the same time. The transport properties of the aggregate are measurable and usually constant in time, while the transport properties of the cement paste depend on the original water/cement ratio, kind and quantity of admixtures, and hydration time, and to some degree on the particle size distribution. However, it will be shown below that the interfacial zone between the cement paste and aggregates may play a critical role in determining the bulk transport properties. Therefore, when averaging over the cement paste microstructure, the micrometre scale interfacial zone, which is determined from cement paste microstructure, must not be left out.

8.6.1 INTERFACIAL ZONE MICROSTRUCTURE

The characteristic features seen in the interfacial zone are: (1) higher capillary porosity than in the bulk and generally bigger pores, and (2) higher CH volume fractions than are seen in the bulk. These features are typically seen in the cement paste volume that is within 50μm of an aggregate surface [8.10].

Using the microstructure model mentioned above, we have analysed two major causes of this interfacial zone microstructure, neither of which depends on bleeding. They are: (1) the particle-packing effect, and (2) the one-sided growth effect [8.11].

The **particle-packing effect** arises from the fact that particles cannot pack together as well near a flat edge as in free space. Since the typical aggregate is many times larger than the typical cement particle, even for the fine aggregate, locally the aggregate edge appears flat to the surrounding cement particles. This inefficient packing causes less cement and higher porosity to be present initially near the aggregate surface, and so even after hydration this condition persists. The width of the interfacial zone will then be in scale with median cement particle size [8.12]. This is the main contribution to the interfacial zone microstructure, but not the only one.

The **one-sided growth effect** arises in the following way. Consider a small region of capillary pore space located out in the bulk paste part of a mortar or concrete. On the average, there is reactive growth coming into this small region from all directions, since the cement particles are originally located randomly and isotropically. Now consider a similar small region of capillary pore space, but located very near an aggregate surface. Reactive growth is coming into this region from the cement side, but not from the aggregate side [8.11].

Mineral admixtures like silica fume and fly ash can also be incorporated into the model, and their effect on interfacial zone properties simulated [8.13]. We have found that the two main variables of importance for mineral admixtures are particle size and reactivity with calcium hydroxide. The size of the mineral admixture controls the width of the packing effect at the aggregate edge, with smaller admixtures allowing better packing nearer to the aggregate edge. The reactivity controls how much calcium hydroxide can be consumed, and converted to CSH. Assuming adequate dispersion, we have found, using the model, that the effectiveness of a mineral admixture in improving the interfacial zone microstructure increases as its reactivity increases and its size decreases [8.13].

Properties of the aggregate can be studied as well, such as the effect of porous and/or reactive aggregates on the interfacial zone microstructure [8.12]. We have found it possible to explain the influence of each of these materials variables on the interfacial zone microstructure in terms of the particle-packing effect and the one-sided growth effect. These ideas then serve as a useful theoretical framework to unify analysis of how material variables affect interfacial zone microstructure [8.12].

8.6.2 CONCRETE CONSIDERED AS A COMPOSITE MATERIAL

If we were to make a concrete out of cement paste and zero porosity aggregate, with no interfacial zones present, then the ionic diffusivity and fluid permeability of the concrete would rigorously have to be lower than the corresponding values for the cement paste. This is because the aggregates, assumed to be fully dense, would have zero transport coefficients, so that the mixture must have lower bulk properties. For this kind of composite, the bulk values of transport coefficients are always less than a simple volume average of the individual phase properties, owing to the random geometry and the nature of the equations that define the various properties, and only decrease as more of the second phase (aggregate) is added. Useful bounds have been derived for the effective bulk properties in terms of the properties of the constituents [8.14].

There are not enough good data in the literature to carefully study the transport properties of concrete using composite theory, which requires simultaneous measurement of the cement paste host and concrete transport properties while the aggregate volume fraction is systematically varied. There is some such careful work, however, on the elastic moduli of mortar [8.15, 8.16]. Some recent work does show that the effective transport properties of concrete can increase greatly as more aggregate is added past a critical amount [8.17]. There are also data showing that concrete can have up to 100 times the water permeability of the cement paste it is made from [8.2]. The only possible microstructural explanation of this behaviour, besides that of extensive microcracking, is the effect of transport of fluid or ions through the interfacial zones. It is already known that the interfacial zone regions contain pores larger than those in the bulk paste. However, if the interfacial zones do not percolate, then their effect on transport will only be negligible, as any transport path through the concrete would have to go the bulk cement paste. Transport properties would then be dominated by the bulk cement paste transport properties.

8.6.3 INTERFACIAL ZONE PERCOLATION

To study the percolation or connectivity of the interfacial zones in concrete is not simple, as the geometry of this phase is complex. Fortunately, in the percolation literature there is a model that is almost perfectly suited for this study: the **hard core/soft shell** model [8.18].

This model starts with a random suspension of 'hard' spherical particles, which are 'hard' in the sense that they are packed without being allowed to overlap, as in a suspension. Then concentric spherical shells are placed around each particle, where the spherical shells all have the same thickness, and are allowed to overlap freely. The volume fraction of shells required to make the shells percolate, which is when a continuous soft shell pathway first becomes established, is then computed. The volume fraction of soft shells required for percolation is a function of how many hard core particles are present, and the thickness of the soft shell. Obviously when more hard core particles are packed in a given volume, there is less space between them, so thinner soft shells will be able to percolate. When there are fewer hard core particles present, thicker shells will be required for percolation of the shell phase. Figure 8.4 shows a schematic drawing of the basic concept of a two-dimensional version of this model.

To study interfacial zone percolation in concrete, we take a fixed shell thickness to represent the interfacial zone, and pack spherical aggregate particles that are then surrounded by these shells. The width of the interfacial zone should be independent of aggregate size, as long as the median aggregate size is at least 5–10 times the median cement particle size [8.12]. The size distribution of the hard core particles is taken from measured aggregate size distributions [8.19–8.21]. The fraction of the total shell volume that forms part of a connected cluster is then computed as a function of the volume fraction of aggregate present.

Results for a mortar are shown in Fig. 8.5, in which each curve shows the connectivity of the interfacial zones for different choices of interfacial zone thickness. When comparing against Portland cement mortar mercury intrusion data [8.19 8.21], it was found that a choice of 20 μm for the interfacial zone thickness gave the best agreement with the mercury data. The mercury data gave an idea of what the percolation threshold of the interfacial zones was by showing a large increase in large pores intruded

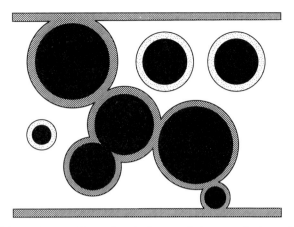

Fig. 8.4. Schematic two-dimensional picture of the percolation of the interfacial zones (grey) around aggregates (black). The darker grey interfacial zones form a percolated cluster.

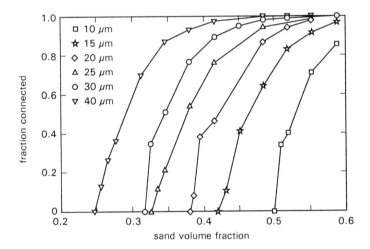

Fig. 8.5. Showing the fraction of the total interfacial zone volume that is a part of a percolated (continuous) cluster as a function of aggregate volume fraction and for several interfacial zone thicknesses.

at the mercury breakthrough point [8.22] at a certain aggregate volume fraction. The width given in scanning electron microscope studies of the interfacial zone, 30–50 μm, is defined by measuring from the aggregate edge to where the measured porosity assumes its bulk value. This would not be the width seen by mercury porosimetry, because it is probable that the larger pores will be found in the larger-porosity region nearer to the aggregate, which will be seen first by the mercury. Also, as to effect on

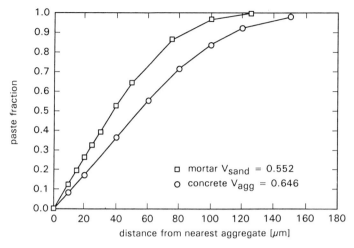

Fig. 8.6. Showing the fraction of the total cement paste volume fraction that lies within a given distance from an aggregate surface, for a mortar with a sand volume fraction of 0.552, and a concrete with an aggregate volume fraction of 0.646.

transport properties, the inner region of the interfacial zone is of more importance, since its transport properties will be higher than the outer region because of the larger pore size and porosity. The width of 20 μm given by the hard core/soft shell model is an effective width, where this width contains the larger pores that would be important for transport. Figure 8.5 also shows that for an aggregate volume fraction of 40% or more and an interfacial zone thickness of at least 20 μm, the interfacial zones will be percolated at least partially, and will be fully percolated for aggregate volume fractions greater than 50%. Most concretes have aggregate volume fractions well above 50%, so that, in general, we can conclude that the interfacial zones in usual Portland cement concrete are percolated, and so will have an effect on transport properties. To calculate the size of this effect requires more work. The fraction of the cement paste that lies in interfacial zones can be calculated with the hard core/soft shell model for a given aggregate volume fraction and particle size distribution. Results are given in Fig. 8.6 for a specific mortar and concrete [8.19]. Figure 8.6 shows that quite a large part of the cement paste matrix lies within an interfacial zone, with about 20% lying within 20 μm, and about 50% lying within 50 μm of an aggregate surface.

8.7 Critically needed future work

In order to further understand the relationships between concrete microstructure and transport properties, more experimental measurements of transport properties must be done on well-characterized samples. To check carefully whether a given analytical theory is useful, or if a computer simulation computation of a transport property is correct, the concrete sample must have the following known quantities: (1) degree of hydration and porosity of the cement paste, (2) the volume and size distribution of aggregates, (3) the width of the interfacial zones, and most importantly, (4) the values of the transport coefficients of interest measured separately for the matrix and aggregate materials.

On the theoretical side, more sophisticated effective medium or homogenization theories need to be developed. It is possible that such effective medium theories, even if made more sophisticated, can never work well on cement-based materials, but the attempt should be made. There is a body of knowledge from homogenization theory and composite materials science that has not been taken advantage of for cement-based materials [8.23]. The simple Maxwell-type theories [8.24] are not enough, since they do not take into account the complexities of cement paste microstructure, nor, when applied to concrete considered as a composite, do they take into account the interfacial zones. More and better fundamental computer simulations [8.25] need to be carried out as well, in order to make microstructure property connections through comparing digital images of microstructure with computed properties. It is also possible that ideas currently being developed in other areas of materials science for **interpenetrating phase composites** (IPCs) will be useful for cement-based materials [8.26]. IPCs have more than one continuous phase, which form interpenetrating networks. Since both the interfacial zones and the bulk cement paste are percolated in concrete, concrete must be considered to be an IPC.

8.8 Conclusions and summary

It has been shown that theoretical understanding of the microstructure transport property relationships of concrete, from the cement paste up to the full composite, can be based on the basic ideas of pore size and connectivity, with connectivity defined rigorously using percolation concepts. These concepts give a fairly complete, although at present mainly qualitative, picture of how transport properties depend on microstructure in concrete. These ideas have already been tested for cement paste and found to be valid. The concept of interfacial zone percolation in concrete needs to be tested, and its effect on transport carefully quantified. Careful materials science experiments, in conjunction with computer simulation computations and new effective medium theories that are based on IPC concepts, can help validate this theoretical understanding of the microstructure–transport property relationships of concrete.

8.9 References

8.1. Allen, A.J., Oberthur, R.C., Pearsons, D., Schofield, P. and Wilding, C.R. (1987) Development of the fine porosity and gel structure of hydrating cement systems. *Philosophical Magazine B*, Vol. 56, No. 3, pp. 263–288.

8.2. Young, J.F. (1988) A review of the pore structure of cement paste and concrete and its influence on permeability, in *Permeability of Concrete*, eds D. Whiting and A. Walitt, ACI SP-108, American Concrete Institute, Detroit.

8.3. Garboczi, E.J., Thorpe, M.F., DeVries, M. and Day, A.R. (1991) Universal conductivity curve for a plane containing random holes. *Physical Review A*, Vol. 43, pp. 64–73.

8.4. Garboczi, E.J. (1990) Permeability, diffusivity, and microstructural parameters: a critical review. *Cement and Concrete Research*, Vol. 20, pp. 591–601.

8.5. Chen, Y. and Odler, I. (1992) On the origin of Portland cement setting. *Cement and Concrete Research*, Vol. 22, pp. 1130–1140.

8.6. Bentz, D.P. and Garboczi, E.J. (1991) Percolation of phases in a three-dimensional cement paste microstructural model. *Cement and Concrete Research*, Vol. 21, pp. 325–344.

8.7. Powers, T.C., Copeland, L.E. and Mann, H.M. (1959) Capillary continuity or discontinuity in cement pastes. *Journal of the Portland Cement Association, Research and Development Laboratories*, Vol. 1, No. 2, May, pp. 38–48.

8.8. Garboczi, E.J. and Bentz, D.P. (1992) Computer simulation of the diffusivity of cement-based materials. *Journal of Materials Science*, Vol. 27, pp. 208–392.

8.9. Christensen, B.J., Mason, T.O., Jennings, H.M., Bentz, D.P. and Garboczi, E.J. (1992) Experimental and computer simulation results for the electrical conductivity of Portland cement paste, in *Advanced Cementitious Systems: Mechanisms and Properties*, ed. F.P. Glasser, G.J. McCarthy, J.F. Young, T.O. Mason, and P.L. Pratt, Materials Research Society, Pittsburgh.

8.10. Scrivener, K.L. (1990) The microstructure of concrete, in *Materials Science of Concrete I*, ed. J. Skalny, American Ceramic Society, Westerville.

8.11. Garboczi E.J. and Bentz, D.P. (1991) Digital simulation of the aggregate paste interfacial zone in concrete. *Journal of Materials Research*, Vol. 6, No. 1, pp. 196–201.

8.12. Bentz, D.P., Garboczi, E.J. and Stutzman, P.E. (1992) Computer modelling of the interfacial zone in concrete, in *Interfaces in Cementitious Composites*, ed. J.C. Maso, E & FN Spon, London, pp. 107–116.

8.13. Bentz D.P. and Garboczi, E.J. (1991) Simulation studies of the effects of mineral admixtures on the cement paste aggregate interfacial zone. *ACI Materials Journal*, Vol. 88, No. 5, pp. 518–529.

8.14. Hashin, Z. (1983) Analysis of composite materials: a survey. *Journal of Applied Mechanics*, Vol. 50, pp. 481–505.

8.15. Zimmerman, R.W., King, M.S. and Monteiro, P.J.M. (1986) *Cement and Concrete Research*, Vol. 16, pp. 239 ff.

8.16. Ulrik Nilsen A. and Monteiro, P.J.M. (1993) Concrete: a three phase material. *Cement and Concrete Research*, Vol. 23, pp. 147–151.

8.17. Houst, Y.F., Sadouki, H. and Wittmann, F.H. (1992) Influence of aggregate concentration on the diffusion of CO_2 and O_2, in *Interfaces in Cementitious Composites*, ed. J.C. Maso, E & FN Spon, London, pp. 279–288.

8.18. Torquato, S. (1986) *J. Chem. Phys.*, Vol. 85, pp. 624–628.

8.19. Winslow, D.N., Cohen, M., Bentz, D.P., Snyder, K.A. and Garboczi, E.J. Percolation and porosity in mortars and concretes. *Cement and Concrete Research*, in press.

8.20. Snyder, K.A., Winslow, D.N., Bentz, D.P. and Garboczi, E.J. (1992) Effects of interfacial zone percolation on cement-based composite transport properties, in *Advanced Cementitious Systems: Mechanisms and Properties*, ed. F.P. Glasser, G.J. McCarthy, J.F. Young, T.O. Mason, and P.L. Pratt, Materials Research Society, Pittsburgh, pp. 265–270.

8.21. Snyder, K.A., Bentz, D.P., Garboczi, E.J. and Winslow, D.N. (1992) Interfacial zone percolation in cement–aggregate composites, in *Interfaces in Cementitious Composites*, ed. J.C. Maso, E & FN Spon, London, pp. 259–268.

8.22. Winslow D.N. and S. Diamond, S. (1970) *ASTM Journal of Materials*, Vol. 5, p. 564.

8.23. McLachan, D.S., Blaszkiewicz, M. and Newnham, R.E. (1990) *Journal of the American Ceramic Society*, Vol. 73, pp. 2187 ff.

8.24. Ross MacDonald, J. (1987) *Impedance Spectroscopy: Emphasizing Solid Materials and Systems*, John Wiley and Sons, New York.

8.25. Garboczi E.J. and Bentz, D.P. (1990) *Materials Science of Concrete Vol. II*, ed. J.P. Skalny, Ed., American Ceramic Society, Westerville.

8.26. Clarke, David R. (1992) Interpenetrating phase composites. *Journal of the American Ceramic Society*, Vol. 75, pp. 739–759.

9 Laboratory test methods

Mette Geiker, Horst Grube, Tang Luping,
Lars-Olof Nilsson and Carmen Andrade

9.1 Steady-state water permeation

9.1.1 INTRODUCTION

Laboratory methods for testing water 'permeability' can be divided into several categories, dependent on the transport mechanism(s) involved:

1. steady-state water permeation;
2. non-steady-state water penetration; and
3. capillary suction (see Chapter 2).

'Permeability' values calculated from data obtained before initial saturation of the sample cannot be interpreted as actual coefficients of permeability, as the assumption of saturated flow is invalid. However, the non-steady-state methods are often used for the purpose of testing, as they are more rapid than most methods of measuring 'true' permeability (steady-state flow). Furthermore, the methods involving capillary suction resemble many of the environmental conditions that concretes are exposed to. The methods involving non-steady-state flow will therefore be dealt with in sections 9.2 and 9.3.

Permeability is expressed as the volume of liquid, per unit area of surface per unit time, flowing through a porous material under a constant pressure head and at a constant temperature; see Chapter 2. The flow is assumed to be one-dimensional. The SI unit for the coefficient of water permeability are $m^3/(m^2s)$ = m/s. In the literature several other units are met. A conversion table for some of these from [9.1] is given in Appendix B. The permeability coefficient of concrete usually varies between 10^{-16} and 10^{-10} m/s.

Performance Criteria for Concrete Durability, edited by J. Kropp and H.K. Hilsdorf.
Published in 1995 by E & FN Spon, London. ISBN 0 419 19880 6.

9.1.2 TEST METHODS

The laboratory methods for testing the rate of water penetration due to a pressure difference (permeation) can be divided into several categories, depending on the property measured and the technique used; see Table 9.1.

Most of the methods are designed for testing concrete. One method deals only with cement paste [9.8] and one is claimed to be applicable to cement paste, mortar and concrete [9.12].

The time to obtain steady-state flow depends on the pressure applied as well as composition (water/cement ratio, degree of hydration, etc.), size and degree of saturation of the sample. The period of time necessary to obtain steady-state flow reported varies from 2 days to 2 weeks ($0.35 < w/c < 0.75$, for vacuum-saturated samples under 3.5 MPa pressure head) to several weeks.

The requirements of the methods are dependent on their purpose. Permeability is measured either for the purpose of research or for testing (acceptance/rejection) of the concrete for a specific project.

A variety of devices for measuring permeation exists. The dimensions of the sample, the magnitude of the applied pressure and the application of a confining pressure to the sample surfaces adjoining the inflow face are some of the parameters varied; see Table 9.1. In a subsequent section various factors affecting the measured permeability are dealt with.

A conventional permeability cell is outlined in Fig. 9.1. Water is pressed through the specimen and collected. No confining pressure is applied. This type of cell has been used for measurement of permeability coefficients down to 3×10^{-11}.

It has been the experience that well-cured materials with low water/cement ratio have permeabilities below the limit of measurements of many types of apparatus [9.20]. Coefficients of permeability as low as 10^{-16} m/s have been determined by means of a recently developed permeability method [9.16, 9.17]. The permeability cell is outlined in Fig. 9.2. The concrete specimen in the form of a disc, 100 or 150 mm diameter and 25–60 mm thick, is contained in a closely fitting cylinder. Before testing, the samples are vacuum-saturated. However, steady-state flow conditions may take several days, perhaps weeks, to establish.

Besides being applicable for measuring permeability in the permeability range of interest there are other requirements for the methods:

Research:
  Low coefficient of variation

Acceptance/rejection:
  Low coefficient of variation
  Rapid
  Applicable on concrete cores
  Preferably simple
  Preferably inexpensive

Table 9.1. Methods of measuring water permeation

Property measured	Specimen Type	Size (cm) (a)	Preconditioning	Duration of test / K in m/s	Pressure applied (MPa)	Confining pressure applied	Coefficient of variation, CV (%) r(d)	R(e)	Reference
Inflow	Concrete	$d=36, h=37.5$	Not specified (b)	ca 14 d	1, 4	None (c)			9.2
	Concrete	$d=10, h=1.3$	23C, 50% RH	14–20 d ($K=5 \times 10^{-12}$)	None	None			9.3
	Concrete	$d=10, h=1.3$	23C, 50% RH	14–20 d ($K=5 \times 10^{-12}$)	None	None	10–15	19	9.3
Outflow	Concrete	$d=10, h=5$	Water sat. (+ 300 Pa)	?	1.0–4.0	None (c)			9.4, 9.5, 9.6
	Concrete	$20 \times 20 \times 20$	Ca 12°C, 50–60% RH	ca 21 d	0.4–0.7	None			9.7
	Paste	$d=1.5, h=0.8$	Water curing, sawing	4–28 d	3 atm	None			9.8
	Concrete	$d=15, h=15$	Not specified (b)	4–14 d	3.5	None			9.4
	Concrete	$d=10, h=20$	Vacuum saturation	2–? d	20	24 MPa	30–70		(f) 9.9
	Concrete	$d=7.6, h=7.6$	Water curing	3 d, $K=10^{-12}$	0.7	1.4 MPa	21		9.10
	Concrete	$d=15, h=5$	Vacuum saturation	ca 60 d, $K=7 \times 10^{-12}$	2.5 bar	None			9.11
	Con./mor./paste	$d=10, h=3.15–17$	Sawing, vac. sat.	ca 2 h, $K=10^{-13}$	0.7–2.8	2.8–14.8			9.12
Inflow and outflow	Concrete	$(d=45, h=45)$	Not specified (b)	8–20 d	2.8	None	25		9.13
	Concrete	$d=16, h=7$	2 d in water, 20°C		1.1	None(c)			9.14
	Concrete	$d=15, h=5$	Not specified	Ca 10 d, $K=^{-12}$	?	None			9.15
	Concrete	$(d=10, h=2.5–6)$	Vacuum saturation	Ca 3 d	1.5–3.5	Yes	10		9.16, 9.17
	Concrete	$d=7.5, h=8$	Vacuum saturation	33 d	0.3	None(c)			9.18

Notes:
(a) d diameter, h height, ( ) other sizes may be used
(b) Examples given in the reference
(c) Cylindrical surfaces sealed
(d) r Repeatability
(e) R Reproducibility
(f) Standard API RP 27

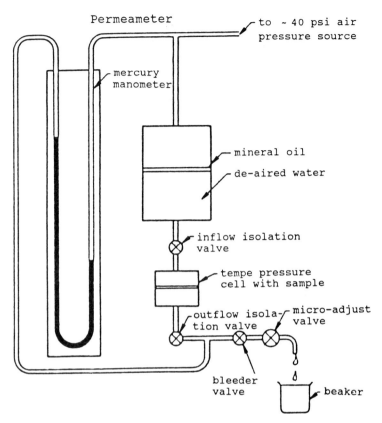

Fig. 9.1. Conventional permeability cell. The cell has been applied for the measurement of coefficients of permeability as low as 3 x 10-11 m/s [9.18].

9.1.3 REPRODUCIBILITY

Only limited information on the reproducibility of the methods appears to exist in the literature. The coefficients of variation obtained are listed in Table 9.1.

9.1.4 STANDARD TEST METHODS

Only one standard recommendation for test of water permeation seems to be available, the American Petroleum Institute's recommendation: API designation RP 27, Recommended Practice for Determining Permeability of Porous Media [9.9].

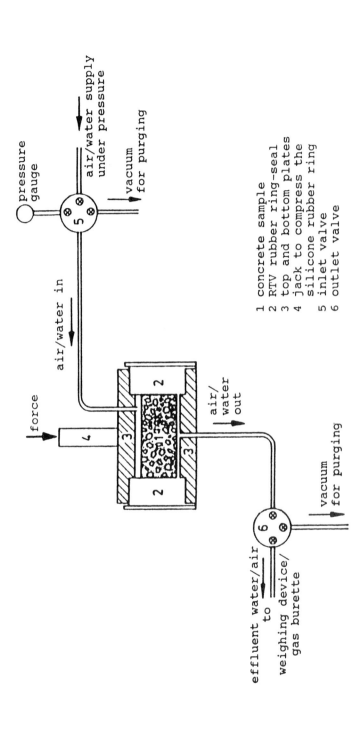

Fig. 9.2. Permeability cell applied for the measurement of coefficients of permeability as low as 10–16 m/s [9.16].

9.1.5 FACTORS AFFECTING THE MEASURED PERMEABILITY

Steady-state flow may be obtained before the sample is saturated, if the humidity on the outflow side is kept below 100%. An error due to the apparent decreased thickness will thus be introduced.

In a saturated sample the water flow in the pores due to hydraulic pressure can be described by D'Arcy's law (see Chapter 2):

$$Q = K_w^* \Delta P \, A / l \qquad (9.1)$$

where Q = water flow (m³/s)
K_w^* = coefficient of water permeability (m/s)
ΔP = pressure difference (m) per water gauge
A = surface area of the sample (m²)
l = thickness of the sample (m)

The permeability of a material is affected by its composition. The water/cement ratio is the most important factor with regard to permeability. The effect of water/cement ratio on permeability was illustrated by Powers [9.19]. Of further importance are the degree of hydration, cement type, content of pozzolanic materials, entrained air, amount of paste, aggregate size and possible defects such as cracks, pinholes and incomplete compaction; see Chapter 4.

When considering the design of a method for testing permeability, various factors that affect the measured permeability should be taken into account. Among these factors are:

(a) Sample size

The effect of the sample size has been dealt with by Hooten and Wakeley [9.12]. Higher values of permeability and a larger scatter of results were obtained for smaller samples. The effect of sample size was negligible for heights of 3 × maximum aggregate size.

(b) Conditioning of the sample

The first drying of cement paste appears to coarsen the porosity significantly [9.20]. This process is irreversible.

Permeability is to be measured on saturated samples. The ease of water saturation is influenced by the permeability of the material (water/cement ratio, degree of hydration, pozzolanas etc.) and the size of the sample. It may not be possible to water-saturate large samples of materials with low permeability. The effect of lack of saturation is dealt with in [9.12] and [9.21]. Compared with a saturated sample, the inflow of water in a non-saturated sample will be higher due to capillary suction, while the outflow of water will be smaller due to the ongoing saturation of the sample.

Especially for incompletely hydrated samples, the duration of the test (water saturation process and permeability test) should be kept at a minimum to minimize the risk of increasing the degree of hydration.

According to Mills [9.22], steady-state conditions may be achieved rapidly if the concrete samples are kept wet (under water) until their permeability is tested.

The water used should be free from air, carbon dioxide and hydrogen carbonate to minimize the risk of air bubbles and carbonation (causing decreasing permeability) and solution of lime (causing increasing permeability, except in concretes that contain blastfurnace slag). According to Gjörv and Löland [9.21] a certain expulsion of dissolved air may take place when the pressure decreases through a concrete specimen. This causes blocking of the pores and decreases the rate of water permeation. By reducing the normal air content of the water by two-thirds, the water permeability of concrete increased by 550% ($w/c = 0.5$) and by 88% ($w/c = 0.6$) [9.21].

(c) Boundary conditions and confining pressure

Unacceptable leakage of water has been found at high driving pressure and low permeabilities [9.18]. The effect of the boundaries depends on, among other things, the ratio between the height and diameter of the samples.

The effect of the boundaries has been minimized by measuring the permeation through the inner part of the sample [9.7, 9.23]. One of the methods measures the permeation through various parts of the sample [9.23].

A few methods take into account the difficulty of sealing the boundaries properly by applying a confining pressure [9.1, 9.10, 9.12, 9.24]. The effect of the ratio between the confining pressure and the driving pressure varies [9.12].

(d) The driving pressure (pressure head)

High pressure may be selected for the purpose of accelerating the test, in order both to obtain data more rapidly and to reduce the risk of continued hydration. High water pressure increases the quantity of water flow available for measurement, thus reducing the error of reading. On the other hand, high pressure increases the risk of altering the pore structure, by silting. The effect of pressure on permeability has been dealt with by Hooten and Wakeley [9.12], who found that increasing driving pressure caused decreased permeability. Glanville [9.25], on the other hand, found the opposite effect during the first few days of the test and later no effect.

(e) Chemical reactions and physical–chemical effects during testing

When water is entering a sample, unhydrated cement will react with the water. The hydrated material fills up the pores, resulting in a decreased permeability.

During the inflow of water into a sample, an osmotic pressure will build up, primarily due to differences in alkali concentration. According to Powers *et al.* [9.8] the osmotic pressure head may be either positive or negative and will in most cases not cause less than 10% deviation in the permeability data obtained.

9.2 Non-steady-state water penetration

9.2.1 INTRODUCTION

The laboratory methods dealt with in this section are methods for testing non-steady-state penetration of water under applied pressure (see Chapter 2 and section 9.1).

9.2.2 TEST METHODS

Laboratory methods for testing non-steady-state water penetration can be divided into two categories depending on the property measured: depth of penetration or inflow (mass/unit area of the specimen).

The difference between the inflow and the outflow is the amount of water absorbed by the specimen. In general, no or only little outflow is obtained during the period of measuring, and the methods may therefore be designated **absorption measurements**.

In Table 9.2 a brief description of various methods is given. The general test set-ups applied when measuring depth of penetration or water inflow are exemplified in Figs 9.3 and 9.4. Water under pressure is applied to one surface of the sample. To measure the depth of penetration, the sample is split perpendicular to the injected face after a specific time, and the depth of penetration is determined visually.

Table 9.2. Methods of measuring non-steady-state water penetration on concrete specimens

Property measured	Specimen size (cm)*	Preconditioning	Duration of test K in m/s	Standard No.	Reference
Depth of penetration	$d=29, h=15$ ($15 \times 15 \times 15$) ($20 \times 20 \times 20$)	Not specified ISO 2735/2 Water curing	24 h 14–20 d ($K=5 \times 10^{-12}$) 14–20 d ($K=5 \times 10^{-12}$)	SS 137214 ISO/DIS 7031 DIN 1048/1	9.26 9.27 9.11, 9.28
Inflow	($15 \times 15 \times 15$) $d=10$ $d=15, h=30$ $d=4–5, h=1$	ISO 2735/2 not specified not specified Dried at 97°C	96 h ca 21 d 2–7 d 8 h	ISO/DIS 7032	9.20, 9.30 9.31 9.32 9.33

* *d* diameter, *h* height, () other sizes may be used

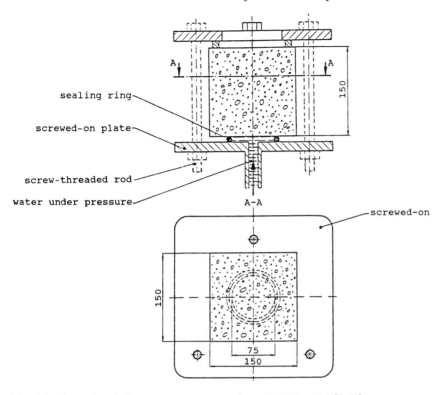

Fig. 9.3. Example of the test arrangements for ISO/DIS 7031 [9.27].

The three methods based on measurement of depth of water penetration, mentioned in Table 9.2, are rather similar. The set-up for the method recommended by RILEM, ISO/DIS 7031, is shown in Fig. 9.3.

Although the results are expressed as depth of penetration, a coefficient of 'permeability' may in some cases be derived. Valenta [9.34] has developed the following equation for uniaxial penetration:

$$d = (2\ K\ t\ h/v)^{1/2} \tag{9.2}$$

where d = depth of penetration at time t
K = coefficient of 'permeability'
h = pressure applied
v = porosity of the sample

The set-up of the RILEM recommended method for measuring water penetration as water inflow, ISO/DIS 7032, is shown in Fig. 9.4. This method has been rejected by the British Standards Institution as the details of the method are not considered to be sufficiently well developed, in

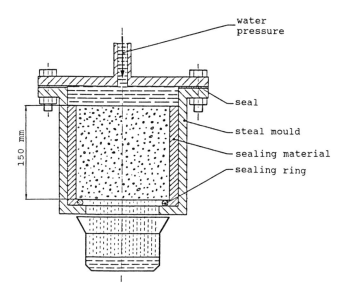

Fig. 9.4. Example of the test arrangements for ISO/DIS 7032 methods [9.30].

particular, with respect to the lack of standard values of applied pressure [9.31]. A similar test technique, but using a fairly sophisticated technique for generating and monitoring constant water pressure, was described by Arnold and Littleton [9.35] and referred to by The Concrete Society, UK [9.31].

Neither of the two remaining methods based on measurement of water inflow appears in its present form to be applicable for general use. The method described by Tyler and Erlin is not offered by the authors as a fully developed method [9.32]. Finally, neutron radiography for measurement of water permeability of concrete as applied by Mo *et al.* [9.33] is not expected to be generally applicable.

9.2.3 REPRODUCIBILITY

No information on the coefficients of variation for the methods listed in Table 9.2 has been obtained, except for the German Standard DIN 1048 [9.28] and the test method proposed by Tyler and Erlin [9.32]. For the latter method at least 50% deviation from a mean of several test results for permeability coefficients at the 2×10^{-12} level seems likely. This is in accordance with data from Bonzel [9.36], who found that the maximum of three results could deviate more than 100% from the mean when using the German method on concretes with $w/c = 0.45$.

9.2.4 STANDARD TEST METHODS

Available standards and recommendations for test of water 'permeability' are listed in Table 9.2. Two of the methods based on measurement of depth of penetration are accepted as standards (SS 137214 and DIN 1048/1); the third is a draft standard (ISO/DIS 7031). None of the methods based on the measurement of the inflow of water during the test is accepted as a standard. One is a draft standard (ISO/DIS 7032).

9.2.5 FACTORS AFFECTING THE MEASURED WATER 'PERMEABILITY'

For discussion of the various factors affecting the measured water 'permeability', refer to section 9.1 and Chapter 4.

9.3 Capillary suction

9.3.1 INTRODUCTION

The laboratory methods dealt with in this section are methods for testing the penetration of water under a negligible applied pressure into a non-saturated sample, i.e. penetration of water due to capillary suction (see Chapter 2 and section 9.1).

9.3.2 TEST METHODS

The laboratory methods for testing water penetration due to capillary suction can be divided into two categories depending on the property measured, weight gain or inflow (mass/unit area of the specimen).

The difference between the inflow and the outflow is the amount of water absorbed by the specimen. As no or only little outflow is obtained during the period of measuring, the methods may be designated **absorption measurements**.

According to Hall [9.37], water absorption into a sample from a water reservoir may be tested by means of three principally different test set-ups (Fig. 9.5): (a) the horizontal case, in which the water absorption rate is independent of gravitational effects; (b) the infiltration case, total flow being the sum of capillary-driven and gravity-driven flows; and (c) the capillary rise case, in which the effects of capillarity and gravity are opposed – a capillary rise equilibrium is ultimately attained. For many building materials, water absorption rates measured in the three configurations are similar because capillarity is dominant. The effect of gravity will not be discussed further.

In Table 9.3 a brief description of various methods is given. The three

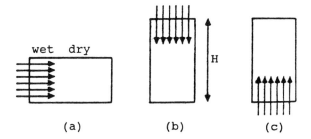

Fig. 9.5. Water absorption into a sample from a water reservoir. Three principally different test set-ups: (a) the horizontal case, in which the water absorption rate is independent of gravitational effects; (b) the infiltration case, total flow being the sum of capillary-driven and gravity-driven flows; (c) the capillary rise case, in which the effects of capillarity and gravity are opposed [9.37].

methods mentioned for measuring capillary suction as weight gain are rather similar. The set-up for a capillary suction (sorptivity) test applied by Hall [9.37] is illustrated in Fig. 9.6. For very accurate work, Hall recommends that the lower parts of the sides of the sample adjoining the inflow face may be sealed with bituminous paint or other coating to prevent absorption. The specimen should rest on rods or pins to allow free access of water to the inflow surface. The water level should not be more than 5 mm above the base of the specimen. The quantity of water absorbed is measured at intervals by weighing the specimen. Surface water on the specimen should be mopped off with a damp tissue. Each weighing should be completed within 30 s, and the clock should not be stopped during weighing. A minimum of five points is necessary to define a good sorptivity plot.

The on-site methods described in section 10.1 (the initial surface absorption test (ISAT) and the Figg method, and variations thereof [9.26]) may of course also be applied in the laboratory. In the laboratory, these

Table 9.3. Methods of measuring water 'permeability' of concrete due to capillary suction

Property measured	Specimen size (cm) [a]	Preconditioning	Duration of test K in m/s	Standard No.	Reference
Weight gain	$(5 \times 5 \times 5)$	Not specified			9.37
	$d=15, h=3$	14 d, 37–47°C	72 h	RILEM 11.2 [b]	9.38
	$15 \times 15 \times 15$	5 d in water, 23 d seal.	22 h		9.26, 9.39
Inflow	$d=5, h=2.5$	Not specified	36–48 h	9.40	

[a] d diameter, h height, () other sizes may be used
[b]   Tentative recommendation

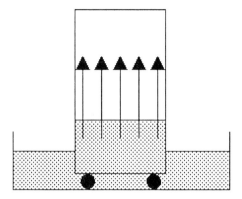

Fig. 9.6. The set-up for the capillary suction test method proposed by Hall [9.37].

methods do not seem to offer any advantages over the test involving uniaxial flow; see Fig. 9.5 [9.37].

The method proposed by Morrison *et al.* [9.40] might turn out to be rather sensitive to room temperature fluctuations due to the small sample size and no pressure being applied and hence the relatively small amount of water absorbed by the sample.

9.3.3 REPRODUCIBILITY

No information on the coefficients of variation for the methods listed in Table 9.3 has been obtained.

9.3.4 STANDARD TEST METHODS

The capillary suction test has, according to Hall [9.37], been adopted in the Swiss Guidelines for Testing, SIA 162/1, Test No 5: 'Water Conductivity' [9.38]. After oven-drying of the test specimens at 50°C for preconditioning the absorption of water after 24 h is recorded for further evaluation.

The method 'Absorption of water by capillarity' [9.41] is a tentative RILEM recommendation (No. 11.2).

9.3.5 FACTORS AFFECTING THE MEASURED CAPILLARY SUCTION

In a recent review paper, Hall [9.37] has dealt with the various factors affecting the measured water absorption.

The cumulative water absorption i increases frequently as the square root of the elapsed time t (see Chapter 2):

$$i = S\sqrt{t}$$

where S is the sorptivity. If the initial water content is not uniform the $\sqrt{t}$ absorption behaviour will not be observed. Deviation from the $\sqrt{t}$ behaviour will also be observed for samples with extremely coarse pore structure exhibiting little suction.

Filling of open porosity on the inflow surface will in general cause the data to show a small, positive intercept on the $\sqrt{t}$ axis. Thus, data are in practice fitted to the equation

$$i = A + S \sqrt{t} \tag{9.4}$$

where A is a positive value.

The rate and amount of water absorption is affected by the initial water content of the sample. This should be kept in mind when relating field and laboratory measurements. When testing cementitious materials in the laboratory, a suitable initial state of dryness should be agreed upon. The effect of moisture and temperature on capillary suction is dealt with in section 4.3.

For discussion of the various factors affecting the measured water 'permeability', refer to section 9.1, and section 4.1 on parameters influencing transport characteristics.

9.4 Gas permeability

9.4.1 INTRODUCTION

The gas permeability of concrete under a differential absolute pressure is a measure of the open porosity prevailing in the concrete; see Chapter 2. Pores that are filled, e.g. with water, are impermeable with respect to this transport mechanism. Therefore, gas flow depends not only on the pore structure of the solids but also varies widely with the moisture content. Consequently, an agreement on the preconditioning of the specimens is necessary to get comparable results.

Moreover, the dimensions of the specimens must be such that the measured value is a real mean value with respect to the inhomogeneity of the concrete. For example, coarse aggregates reaching from one surface of the specimen to the other may cause singularities and increase the scattering of the results.

As a materials coefficient K_g (m^2), according to eq. (2.6) in Chapter 2, is determined for a steady-state laminar flow, taking into account the compressibility of the gas [9.42–44]. Non-steady-state flow combined with a partial vacuum is used in [9.45]. When neglecting the compressibility of the gas in the specimen a coefficient K_g (m²/s) according to eq. (2.7) in Chapter 2 can be calculated.

Testing gas permeability is a rather rapid procedure: in general the time needed for one test is less than an hour.

The gases used should not react chemically with the concrete, in order to avoid a change of the pore structure during the test. Recommended testing gases are oxygen (O_2) or nitrogen (N_2) but not e.g. carbon dioxide (CO_2) [9.46].

9.4.2 TEST METHODS

A uniaxial gas flow between the two parallel surfaces of the test specimen is caused by a different absolute pressure of the test gas on both surfaces; see Fig. 9.7. The flow depends on the pressure difference, testing

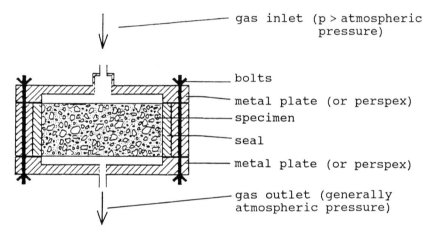

Fig. 9.7. Principle of measuring gas permeation with steady-state flow.

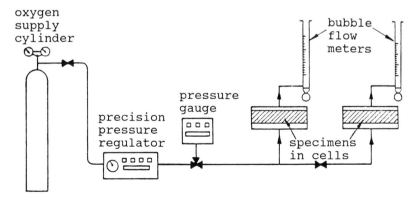

Fig. 9.8. Line diagram for the oxygen permeability apparatus (steady-state flow).

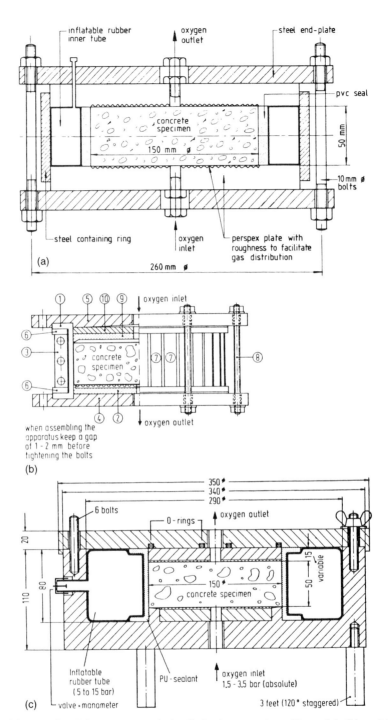

Fig. 9.9. Details of the recommended cells in three versions. Types (a), (b) and (c). Key to type (b): (1) sealant (PUR), Shore hardness scale A-55; (2, 9, 10) supporting plates, oxygen distributors (hard PVC); (3) clamp (split ring, 2 mm sheet steel, two halves closed by bolts); (4, 5) bottom and top plates (steel); (6) load distributors (hard PVC, four half rings); (7) supports (steel, 30 pieces); (8) bolts (steel, 6 pieces)

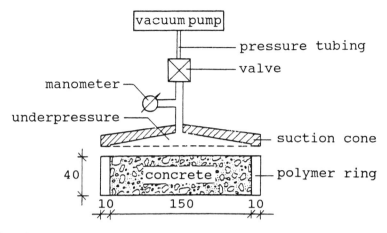

Fig. 9.10. Principle of measuring gas permeation with non-steady-state flow [9.42].

area, thickness and open porosity of the specimen and the viscosity of the test gas. The testing equipment consists of a gas supply, a pressure regulator with pressure gauge, the testing cell(s) and flow meter(s) (Fig. 9.8). Figure 9.9 shows different types of testing cells [9.42], which may be distinguished because of different seals or handling techniques used. The coefficient of permeability K_g (m²) is calculated according to eq. 2.6, Chapter 2.

In another test [9.45] the specimen is cast in a ring of rubber on which the testing equipment (Fig. 9.10) can be mounted. The sealing is achieved by the negative pressure caused by the vacuum pump. After the valve between the vacuum pump and the testing cell is closed the development of pressure rise in the testing cell is determined. The adjustment of the inside pressure is caused by a non-steady-state flow of air through the specimen. Neglecting the compressibility of the gas, a coefficient of permeability K_g is calculated in [9.45] with the dimensions m²/s (see Chapter 2, eq. (2.7)).

9.4.3 PERFORMANCE OF THE TEST

(a) Pressure, gases, duration of the test

The difference of pressure between both surfaces should be as small as possible [9.42]. Pressures at the inlet between 0.5 and 2.5 bar above atmospheric pressure are generally used. Oxygen and nitrogen are well-tried test gases [9.42, 9.47, 9.48]. In the procedure according to [9.45] air is used as a testing gas.

The lower the permeability of a specimen, the longer is the time needed to get a constant gas flow. Therefore, a sufficient time for adjustment, e.g. > 30 s [9.42], is necessary. The vacuum procedure [9.45] needs a longer time for adjustment.

(b) Test specimens

Specimens according to [9.42] have a diameter of 150 mm and height of 50 mm. Tests show that thinner specimens lead to larger permeability co-efficients [9.45, 9.48], all other parameters kept constant. For laboratory tests on specimens prepared in moulds two specimens for each parameter are recommended as a minimum [9.42, 9.48]. The use of more specimens for one parameter may be suitable; see section 9.4.4.

(c) Preconditioning of specimens

Measuring the open porosity of concrete with gas requires that the main pores are filled with gas (air) and not with water. Therefore, it is necessary to precondition the specimens in constant climate prior to the permeability test; see section 4.3.

Two types of preconditioning for example are mentioned in [9.42]:

A: $20 \pm 2°C$ and $65 \pm 5\%$ RH for 28 days
B: $105 \pm 5°C$ for 7 days followed by $20 \pm 2°C$ and 0% RH for 3 days.

Type A simulates normally encountered exposure. Type B results in substantially higher permeabilities. The results of permeability measurements can be compared only if the procedure and the preconditioning are the same. Then, different permeabilities, e.g. due to different water/cement ratios or different curing durations, can be distinguished. Numerous other types of preconditioning may be suitable, but a restriction is recommended with respect to comparability of the results; see section 4.3.

In order to get constant drying conditions for the circular testing surfaces of the specimens it is recommended that the cylindrical area is sealed with an elastic rubber band or a coating during preconditioning.

9.4.4 EVALUATION OF THE RESULTS, VARIABILITY

The range of permeability coefficients K_g for dense concretes of strength classes of 15–55 MPa is from about 10^{-14} to 10^{-19} m^2 [9.42, 9.47]. Calculated values for Kg normally decrease if higher pressures are applied, because the flow may not be strictly laminar any more.

If high-precision manometers are used and the sealing of the specimens is perfect a remarkable precision of the coefficient of permeability can be achieved. The individual results of five tests performed by the same operator in the same laboratory on the same specimen should not show a coefficient of variation of more than 2% [9.42].

Looking at a series of 10 specimens cast from the same concrete batch and cured in the same way, the single-operator coefficient of variation has been found to be about 30%. Taking into consideration the five orders of magnitude for different concretes, the mean value of only two specimens is already highly characteristic.

9.5 Gas diffusion

9.5.1 INTRODUCTION

The transfer of gases by diffusion is stimulated by a difference in concentration, i.e. a difference in partial pressure of the gases and not in absolute pressure (see Chapter 2). The objective of the test is to measure the diffusion coefficient D (m^2/s) in Fick's law according to eq. (2.1) in Chapter 2.

The diffusion coefficient depends not only on the open porosity of the concrete, because diffusion of gases is also possible in pores that are filled with water. Moreover, the type of gas influences the diffusion coefficient. There are considerable differences between gases that do not react with the concrete like oxygen and nitrogen, and gases that can condense and/or react with the surface of the hardened cement paste.

Water vapour is concerned in either respect. Therefore, it is called **diffusion coefficient of water vapour** [9.49]. As hydrated cement paste (hcp) reacts with water, thereby changing its permeability, the tests may need a lot of time. The testing conditions, the partial pressures and the temperature are extremely important for evaluating the result. Corresponding behaviour should be observed for vapours of organic liquids [9.50], which may condense, or for carbon dioxide, which reacts chemically with calcium hydroxide in the hcp [9.46].

9.5.2 TEST METHODS

In gravimetric procedures the specimen is used as a cap of a leakproof tank [9.49]; see Fig. 9.11. Defined but different partial pressures inside and outside the tank may cause mass changes of the equipment (tank +

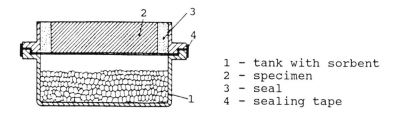

1 - tank with sorbent
2 - specimen
3 - seal
4 - sealing tape

Fig. 9.11. Model for gravimetric diffusion test [9.49].

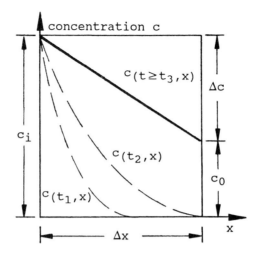

Fig. 9.12. Diffusion changing from non-steady state to steady state.

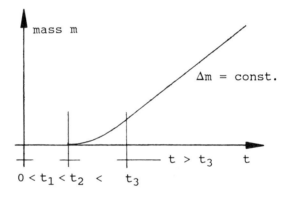

Fig. 9.13. Change of mass during gravimetric diffusion test at non-steady- and steady-state flow.

specimen + content). The latter may be a sorbent or a vaporizing liquid. Diffusion being initiated needs a lot of time (t_1, t_2) until steady-state diffusion is reached. The constant loss of mass (or the constant rise of mass, see Figs 9.12 and 9.13) is a measure of the diffusion coefficient. Measurements are carried on until a constant rate of weight change is reached with respect to the precision of weighing (steady-state diffusion).

The testing method of Schwiete and Ludwig [9.51] using the test gases oxygen and nitrogen (Fig. 9.14) is well suited to obtaining the diffusion coefficient of oxygen for concrete specimens of cylindrical shape. Lawrence [9.48] described a similar method but he added two flowmeters at the outlets to get a better control (Fig. 9.15). The testing cell in Fig.

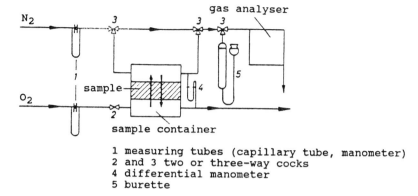

Fig. 9.14. Scheme of testing equipment for diffusion tests with defined continuous gas flow [9.51].

1 measuring tubes (capillary tube, manometer)
2 and 3 two or three-way cocks
4 differential manometer
5 burette

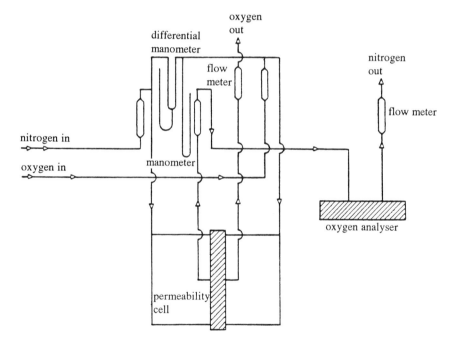

Fig. 9.15. Scheme of testing equipment for diffusion tests with defined continuous gas flow with flow meters at the outlets [9.48].

9.14(c) is equally suitable for diffusion measurements when omitting the bearing plates in the cell and installing a gas inlet and outlet at both surfaces of the specimen.

The evaluation of the diffusion coefficient is based on Fick's first law, eq. (2.1), Chapter 2. The transmission rate i (kg/m^{2}s) of oxygen is determined with a gas analyser as an impurity occurring in the nitrogen due to the oxygen flow through the specimen. Equations for calculating the diffusion coefficient D (m^2/s) are specified in [9.48, 9.51, 9.52].

Oxygen and nitrogen have proved to be cheap and nearly inert testing gases for concrete [9.48]. The testing time may be hours or days until a constant gas flow is achieved, depending on the tightness of the specimen.

Another testing method is described in [9.53, 9.54] using a non-steady-state flow (Fig. 9.15). After the testing cell is rinsed with nitrogen, diffusing oxygen from the open air will accumulate in the cell. Its concentration is measured continuously. The evaluation of the diffusion coefficient is based on Fick's second law; see Chapter 2.

9.5.3 PERFORMANCE OF THE TEST

(a) Test specimens; duration of the test

Concrete test specimens used had diameters of 110–150 mm and thicknesses of 20–50 mm. Preconditioning is necessary as described earlier; see section 9.4. Microcracks should not reach through the specimen. The thickness should be adjusted to the maximum grain size of the aggregate, which means 1–3 times maximum grain size [9.42, 9.55]. If this relation cannot be met, e.g. for drilled cores, the greater scatter of results may be compensated by a larger number of specimens. For laboratory tests on specimens cast in moulds a minimum of two specimens for each parameter is recommended [9.42, 9.55] with diameter of 150 mm and thickness of 50 mm. The use of more specimens for one parameter may be suitable; see sections 9.4.4 and 9.5.4. Several hours may be needed to get a constant diffusion.

(b) Testing equipment; testing cell

Small tanks of glass or metal (Fig. 9.15) are used for the gravimetric procedure according to section 9.5.2. The specimens sealed on the cylindrical surface are used as a cap of the tanks [9.49]. For the continuous steady-state procedure [9.42, 9.48, 9.51] according to section 9.5.2, testing cells as described in section 9.4.2 may be used. The volume of gas on both free surfaces of the specimens should be big enough to avoid insufficient exchange of the gases.

9.5.4 EVALUATION OF THE RESULTS; VARIABILITY

The range of gas diffusion coefficients for dense concretes of strength classes of 15–55 MPa is from about $D = 10^{-6}$ m^2/s to 10^{-10} m^2/s [9.48].

Presumably the high amount of effort for the diffusion tests is the reason for a lack of tests investigating the variability of the results.

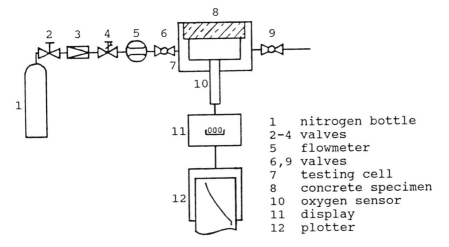

1	nitrogen bottle
2-4	valves
5	flowmeter
6,9	valves
7	testing cell
8	concrete specimen
10	oxygen sensor
11	display
12	plotter

Fig. 9.16. Testing equipment for the sensor method [9.53, 9.54].

9.6 Transport of ions: ion diffusion in concentration gradients

9.6.1 INTRODUCTION

Transport of ions in concrete, especially chloride ions, plays an important role in the durability of a reinforced concrete structure. Ion transport in concrete is a rather complicated process, which involves diffusion, capillary suction and convective flow with flowing water, accompanied by physical and chemical binding. These complications are usually neglected, and pure diffusion, sometimes with binding capacity, is adopted as a predominant transport process in the evaluation of the ion transport property. There are many methods reported for evaluating ion diffusion in hardened cement paste, mortar and concrete. They can be classified as three kinds: steady-state, non-steady-state, and electrical methods.

9.6.2 STEADY-STATE METHOD

The steady-state method is generally referred to a diffusion cell experiment. This method involves: (1) arranging a thin slice of specimen (1 mm to 15 mm thick) to separate two reservoirs (cells) of solution, one (cell 1) containing specific ions (diffusing ions) and the other (cell 2) not containing them; and (2) monitoring the increase of the specific ions in the solution of the latter cell in a certain interval of time. On the assumption that Fick's first law is applicable after steady-state flow of the ions is reached, the diffusion coefficient can be directly calculated from the experimental data by using the following equation:

$$D = \frac{V \Delta Q}{A \Delta t} \times \frac{L}{(c_1 - c_2)}$$

(9.5)

where
V = volume of cell 2 (m³)
ΔQ = increase of the concentration of the specific ions in cell 2 (kg/m³)
Δt = time interval (s)
A = area of slice exposed to the solution (m²)
L = thickness of slice (m)
c_1 = concentration of the specific ions in cell 1 (kg/m³)
c_2 = average concentration of the specific ions in cell 2 (kg/m³).

It is of extreme importance to ensure the separation of two cells by the slice to be tested, otherwise the result will be completely wrong due to the leakage of solution. The time required to achieve a steady-state condition may be weeks or months, depending on the properties and the thickness of the specimen.

The steady-state method was first reported by Gordon [9.56]. Ushiyama and Goto [9.57] first published a detailed study on the diffusion of various ions in hardened cement paste by using this method. A typical experimental arrangement is shown in Fig. 9.17. A saturated calcium hydroxide solution in both cells is usually considered to prevent lime leaching from the hardened cement paste. Page *et al.* [9.58] reported a detailed procedure for a chloride diffusion experiment. The cement paste was cast into moulds, Ø 49 × 75 mm, which then were subjected to continuous rotation of 8 rpm for 48 h around a horizontal axis to prevent significant segregation. Then the specimens were demoulded and cured in a saturated calcium hydroxide solution at 22°C for 60 ± 3 days. The disc sample,

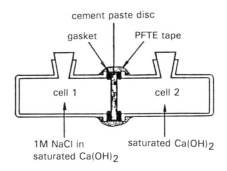

Fig. 9.17. Typical diffusion cells [9.58].

thickness about 3 mm, was cut from the central region of the specimen and its surfaces were slightly ground on grade 600 emery paper, rinsed with distilled water and dried with tissue before being fitted into the diffusion cell shown in Fig. 9.17. The concentrations of chloride ions in cell 2 were determined by withdrawing 100 μl aliquot of the solution and analysing them after various diffusion times. The volumes of the solution withdrawn were neglected in the subsequent calculation as compared with the whole volume of the solution in cell 2.

Roy *et al.* [9.6.4] developed a simpler type of diffusion cell as shown in Fig. 9.18. The sample cement paste disc, ∅35 × 1.68 mm or ∅35 × 3.18 mm, was directly moulded in a hollow Teflon ring mould as shown in the left side of Fig. 9.18, and cured in saturated calcium hydroxide solution for the required time. The disc was sealed onto the rim of the diffusion cell, a 3.5 cm outer diameter plastic petri dish, as shown in the right side of Fig. 9.18, in a bath filled with saturated calcium hydroxide solution so as to keep the dish full of solution, and a thin layer of an underwater-setting epoxy and a heat-shrink moisture-proofing tape was applied onto the sides of the dish and the disc. The cell was immersed in a 0.05 M caesium chloride solution saturated with respect to calcium hydroxide. After a specified time, the cell was removed. The disc was prised open and the entire quantity of liquid in the cell was removed for analysis. This method is relatively simple, although it also needs a long testing period. However, it is not certain that the concentration of chloride ions diffusing into the cell is negligible as compared with the concentration of the solution outside the cell.

Hansson *et al.* [9.60] proposed another simpler type of diffusion cell as shown in Fig. 9.19. The 3 mm thick disc of hardened cement paste was glued to the neck of a plastic bottle; then the bottle was inverted, set on a plastic support in a tank containing a 1 M NaCl solution saturated in $Ca(OH)_2$, and filled to the level of the tank solution with a half-saturated $Ca(OH)_2$ solution. The amount of chloride diffusing through the disc was

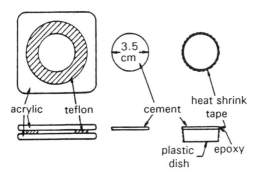

Fig. 9.18. Roy *et al.*'s diffusion cell [9.59].

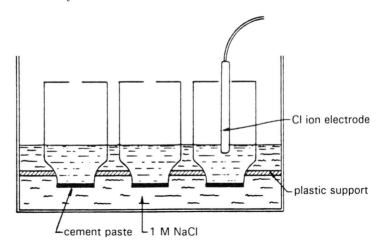

Fig. 9.19. Hansson *et al.*'s diffusion cells [9.59].

determined by periodically measuring the chloride content of the solution in the bottle using a chloride ion selective electrode. The advantage of this type of cell is that several bottles can be set in the same tank.

9.6.3 NON-STEADY-STATE METHOD

The non-steady-state method is generally referred to as an **immersion experiment**. This method usually involves: (1) sealing all except one surface of the specimen to prevent multi-directional penetration; (2) immersing the specimen in a solution containing specific ions for a certain time; and (3) measuring the penetration depth or penetration profile of the specific ions in the specimen. On the assumption that Fick's second law of diffusion is applicable, the diffusion coefficient can be calculated from the experimental data by using eq. (9.6) for penetration depth [9.61] or eq. (9.7) for a penetration profile [9.62]:

$$x_d = 4\sqrt{Dt} \tag{9.6}$$

$$c = c_0\left[1 - \mathrm{erf}\left(\frac{x}{2\sqrt{Dt}}\right)\right] \tag{9.7}$$

where x_d = penetration depth (m)
t = duration of immersion (s)
c = ion concentration at the distance x (kg/m^3)
c_0 = ion concentration at the exposed surface (kg/m^3)
erf = error function

x = distance from the exposed surface (m)

The initial condition of the specimen is very important. If the pore system in the specimen is not saturated with water, capillary suction and moisture transport will occur. In this case, the assumption of pure diffusion is manifestly inappropriate. Therefore, the specimen should be well saturated with water before immersion. Pure immersion experiments are rather time-consuming. The immersion period may be at least months, or even years, mainly depending on the properties of the specimen.

The penetration depth is usually determined by means of a colorimetric method [9.61], and penetration profiles can be obtained by the following procedure: sampling the specimen at different depths by cutting or drilling successively from the exposed end, then analysing the amount of the specific ions in the samples by general chemical analysis or modern techniques including fluorescent X-ray spectroscopy and energy-dispersive X-ray analysis. A special grinding technique has been developed that allows samples to be taken from layers down to 0.5 mm of thickness of the specimen [9.63].

With the development of personal computers, the diffusion coefficient D nowadays can be easily obtained by sophisticated curve-fitting of eq. (3) to penetration profiles. Poulsen [9.64] also proposed a method to calculate D from penetration profiles by using a portable calculator.

So far, the only standardized test procedure of an immersion experiment is AASHTO Designation T 259-80 [9.65]. The specimen of 300 × 300 × 75 mm concrete slab is ponded with a 3% NaCl solution for a period of 90 days, then the solution was removed and the slab is sampled at various depths using a 25 mm diameter rotary hammer drill. The total chloride ion content of each sample is analysed by using nitric acid digestion and a potentiometric titration procedure.

9.7 Transport of ions: electrical methods – theoretical background

9.7.1 INTRODUCTION

When a concrete is completely water saturated, chlorides may penetrate by a pure diffusion mechanism, which has been extensively discussed in other chapters of this report. In this section, the effect of a superimposed electrical field will be discussed, as well as its potential to determine the diffusion coefficient D in an accelerated test in stationary and non-stationary conditions.

Earlier experiments showed that chlorides move quicker through concrete when an electrical field is applied [9.66, 9.67], and corresponding methods are used at present for chloride removal [9.68, 9.69]. In fact, this

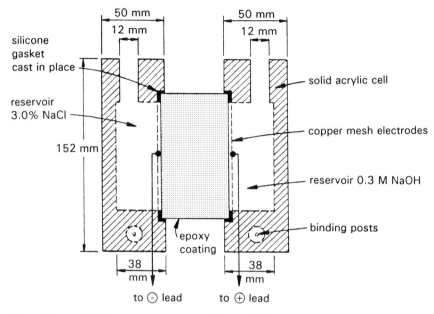

Fig. 9.20. AASHTO test arrangement [9.70, 9.71].

chloride migration was already experienced by many researchers using electrochemical techniques (as cathodic protection for instance). However, it was Whiting [9.70, 9.71] who proposed a **rapid chloride permeability test** in order to obtain an appraisal on concrete permeability within a few hours. This standard test has promoted a strong controversy [9.72, 9.73] on the potential to predict the resistance of concrete to the permeation of chlorides [9.74]. Nevertheless, the test is increasingly used, although everybody recognizes the still existing uncertainties involved.

The test uses a thick (usually 5 cm) concrete disc between two electrodes (usually copper meshes) in an arrangement similar to that of the diffusion cell. Sodium chloride (3% per weight) is added to one of the chambers and NaOH of about 0.1 M to the other. Then, an electrical field of 60 V is applied between the electrodes and the amount of coulombs are recorded during 6 h of testing. The test defines that a higher amount of coulombs represents a higher permeability of the concrete for chlorides. Figure 9.20 shows the arrangement suggested by Whiting.

9.7.2 PROCESSES IN CONCRETE SUBMITTED TO AN ELECTRICAL FIELD

When an electrical field is applied to an arrangement as shown in Fig. 9.21, the processes developing are those also occurring in liquid electrolytes, as follows:

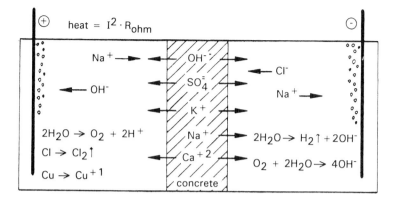

Fig. 9.21. Processes occurring when an electrical field is applied in a diffusion cell: Joule effect, anode dissolution, electrolysis of the electrolyte (gas evolution at electrodes and reduction reactions) and ionic migration and diffusion.

(a) Electrode processes

The current applied to the cell is spent in the electrode processes, by either

- metal dissolution, or
- electrolyte reduction or oxidation [9.75–9.77].

If the anode is an oxidizable metal, it dissolves, thereby generating soluble species and oxides. In the case of graphite electrodes, the process will be

$$C \rightarrow CO + CO_2$$

Gases may be generated from the electrolyte due to its oxidation/reduction. For instance, water electrolysis will generate H_2 and O_2 evolution. Also, other species may generate gases, e.g. $Cl^- \rightarrow Cl_2$ or the carbon oxidation already mentioned.

(b) Movement of ions in the electrolyte

The current applied provokes the movement of ions across the electrolyte. This phenomenon is known as **migration**.

The proportion of the current carried by an ion with respect to the total current is defined as the **transference number** [9.80, 9.81], which is a function of the equivalent conductivity:

$$t_j = \frac{i_j}{i} = \frac{\lambda_j}{\Lambda}$$

$$(9.8)$$

(c) Joule effect

If the current is high enough, heat can be generated [9.80, 9.81], and an increase in temperature is noticed. This increase in temperature will affect the mobility of ions.

Considering these effects it can be deduced that the rapid chloride permeability test contains the following errors:

- The total current is measured and not that portion that corresponds to the chloride flux.
- When integrating the total current from the beginning of the experiment, it does not distinguish between chloride flow plus reaction or simple flow.
- The high voltage drop used (60 V) induces heat, which in turn changes the flow speed.

Therefore, a migration test of this type cannot provide information on the transport of chlorides, nor on the porosity or 'permeability' of the concrete specimen.

9.7.3 DIFFUSION AND MIGRATION PHENOMENA

Ion diffusion in an electrolyte is governed by its mobility, u_j, a basic parameter linked to ion radius, charge and degree of solvation [9.78–9.80]. The ion mobility is usually expressed in cm^2/sV, and the standard values are calculated for infinite dilution; increasing concentrations will reduce mobility due to ion–ion and ion–solvent interaction.

The proportionality factor relating ionic charge Z_j, ionic conductivity λ_j, mobility U_j and concentration C_j is the Faraday constant:

$$\lambda_j = F Z_j U_j C_j \tag{9.9}$$

To characterize diffusion processes, the main parameter used is the diffusion coefficient D, according to the empirical Fick's laws, linking the amount of ions moving per unit of time and cross-section (ionic flow) J, with the gradient of ion concentration dc/dx. Assuming stationary conditions:

$$J = D \frac{dc}{dx} \tag{9.10}$$

the diffusivity D is a function of the ionic mobility:

$$D = U_{abs} K T \tag{9.11}$$

where U_{abs} = absolute mobility
 K = Boltzman constant
 T = absolute temperature

9.7.4 NERNST–PLANK EQUATION: STATIONARY CONDITIONS

Diffusion can be accelerated by an electrical field. This acceleration is called **migration,** and is governed by the same basic parameter as diffusion, i.e. by the ionic mobility (diffusivity) D.

The equation that models general ionic movements in electrolytes is given by **Nernst–Plank**, which can be written for unidirectional movements in stationary conditions as

$$-J(x) = D\frac{\partial C_j(x)}{\partial x} + \frac{ZF}{RT}D_jC\frac{\partial Ex}{\partial x} + C_jV_jx \tag{9.12}$$

Total flux = diffusion + migration + convection

where $J(x)$ = unidirectional flux of species j (mol/cm^{2}s)
 D_j = diffusion coefficient of species j (cm^2/s)
 ∂C = variation of concentration or activity (mol/cm^3)
 ∂x = variation of distance (cm)
 Z_j = electrical charge of species j
 F = Faraday's number (C/eq)
 R = gas constant (cal/V eq)
 T = absolute temperature (K)
 C_j = bulk concentration of the species j (mol/cm^3)
 ∂E = variation of potential (V)
 V = artificial or forced velocity of ion (cm/s)

A rigorous solution for this equation cannot be obtained for solutions as concentrated as the concrete pore solution, which is a polyelectrolyte of elevated ionic strength. However, if only an approximate or average value is enough, the introduction of some simplifying assumptions will allow the calculation of the ion diffusivity D.

The main simplifying assumptions needed are two:

1. No convection is produced inside the concrete.
2. Diffusion is negligible compared with migration when the external electrical field applied is higher then 10 V.

Thus, the Nernst–Plank equation can be condensed to:

$$J = \frac{ZF}{RT} D_j C_j \frac{\Delta E}{x}$$

(9.13)

and so D can easily be calculated provided stationary conditions are established in a traditional diffusion cell.

In stationary conditions an 'effective' value of D, D_{eff}, is obtained: i.e. chloride reactions with the cement paste phases are not taken into account. This equation for stationary conditions may be also written as:

$$D_{eff} = \frac{J_{Cl}RTx}{ZFC_j\Delta E}$$

(9.14)

in which all parameters are known and J_{Cl} can be calculated from an experimental test in which the amount of chlorides is monitored versus time. A graphical presentation of the Nernst–Plank equation is given in Fig. 9.22.

9.7.5 CALCULATION OF D FROM THE VALUE OF INTENSITY: NERNST–EINSTEIN EQUATION

The diffusion coefficient D may also be calculated from recording of the intensity i during the experiment, because it is well established [9.76, 9.78, 9.80] that the flux of a migrating species is proportional to the total intensity:

$$J_{Cl} = \frac{it_j}{nF}$$

(9.15)

The transference number (of the chloride in this particular case) repre-

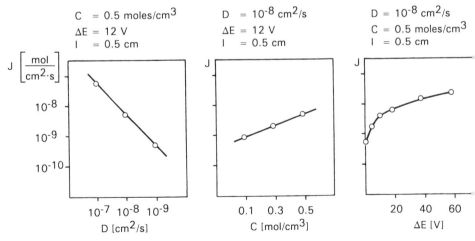

Fig. 9.22. Chloride flux J according to the Nernst–Plank equation [9.71] (migration term) as a function of chloride activity values.

sents the proportionality factor. Therefore, it is possible to calculate the transference number t_j from the value of J_{Cl} obtained in an experiment.

By substitution of eq. (9.15) in the Nernst–Plank equation (9.13), the following expression is obtained:

$$D_j = \frac{RT}{nF^2} \Lambda_j = \frac{RT}{nF^2} \cdot \frac{it_j}{\Delta E} \cdot \frac{x}{A} \cdot \frac{1}{C_j Z}$$

(9.16)

where A =   cross-section of the concrete disc

This equation, denoted as the **Nernst–Einstein equation**, offers the possibility of calculating the diffusion coefficient D from a simple measurement of the resistivity or the conductivity provided that t_j of the particular ion and its concentration are accurately known.

Figure 9.23 shows the graph of the Nernst–Einstein equation (9.16) assuming a solution 0.2 M NaOH plus NaCl with activity values of 0.1, 0.35 and 0.5 mol/cm³ and a chloride transference number of 0.338.

It can be deduced from this figure that D varies with the concentration as theoretically stated [9.76]. The resistivity represents the most influencing parameter.

9.7.6 EXPERIMENTAL PROCEDURES FOR THE DETERMINATION OF DIFFUSIVITIES IN STATIONARY REGIMES

In order to calculate the effective chloride diffusion coefficient for concrete in a stationary regime there are three possibilities [9.81]:

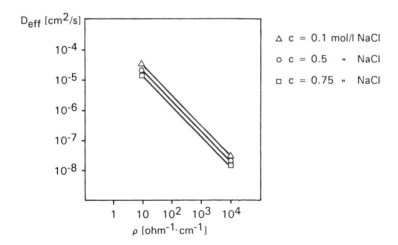

Fig. 9.23. Graph of Nernst–Einstein equation (9.16) as a function of chloride activity values.

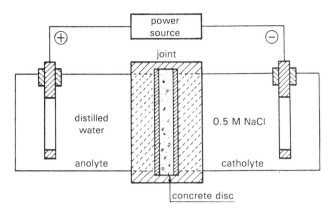

Fig. 9.24. Migration cell.

1. from an arrangement similar to a 'diffusion cell', applying a constant voltage and recording chloride variation in the anolyte versus time;
2. from the same arrangement but in addition to chloride flux recording (or applying) the current flowing through the cell;
3. from resistivity values of concrete specimens.

(a) Constant voltage: Nernst–Plank equation

The cell may be of any adequate size similar to that described in ASTM [9.74] or as shown in Fig. 9.24.

The thickness of the concrete disc has to be not too thick (between 5 and 10 mm) in order to facilitate a quick achievement of the steady-state conditions.

As catholyte, 0.5 M NaCl is suitable. Any concentration is possible, although it is important to remember that activity and not concentration must be used in the equations.

As anolyte, distilled water is the most appropriate in order to avoid undesirable Cl_2 evolution, which can be produced if alkaline solutions, e.g. NaOH, are used.

As electrodes, steel or any corrodable metal may be used as anodic electrode. A non-corrodable anode will produce too high a water electrolysis or Cl_2 evolution. As cathode, steel can also be used, because H_2 evolution can never be avoided there, unless a Ag/AgCl electrode is used.

The voltage drop has to be higher than 10 V in order to accelerate the process, but not high enough to produce heat. A voltage of 12–15 V is appropriate.

Typical results for the chloride flux obtained in corresponding tests are shown in Fig. 9.25, where the chloride content in the anolyte is plotted versus time, after applying a constant voltage of 12 V.

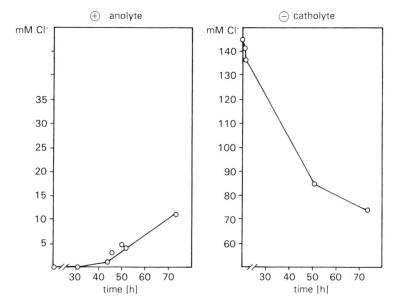

Fig. 9.25. Chloride concentration in anolyte and catholyte in the migration cell versus time.

The flux of chlorides was calculated from the slope of the straight part of the plot shown in Fig. 9.25 as 3.27×10^{-9} mol/cm^{2}s. Introducing this value in eq. (9.14) the resulting D_{eff} value was 1.49×10^{-8} cm^2/s.

(b) Constant voltage or current: Nernst–Einstein equation

When the intensity is recorded together with the chloride flow, or a galvanostatic test is provided, it is necessary to calculate the transference number from eq. (9.15) first. In the present case the intensity recorded was about 24 mA, and therefore the transference number $t_{Cl} = 0.37$, which is very reasonable.

Applying now eq. (9.16), the D_{eff} value resulted in $D_{eff} = 3.44 \times 10^{-8}$ cm^2/s, not far from that calculated previously.

(c) Calculation from resistivity values

Equation (9.16) can also be written as:

$$D_{eff} = \frac{RT}{nF^2} \cdot \frac{t_{Cl}}{\rho c \gamma} \tag{9.17}$$

which allows the calculation of D_{eff} from a simple, although accurate, measurement of concrete resistivity, provided that the chloride activity, c, is known or assumed. Two ways are feasible:

1. If the concrete specimen under investigation already contains chlorides, their proportion in the pore solution, or at least, the water-soluble chloride content, is needed in order to calculate c.

2. If the concrete specimen does not contain chlorides, a value of c between 0.3 and 0.5 can be used, because the resistivity of alkaline pore solutions decreases by 2–3 times if chlorides are added in concentrations of 0.1–1 M.

Although an approximation, this last procedure has a very wide practical application, because it allows the calculation of D_{eff} from a simple resistivity measurement.

9.8 Transport of ions: electrical methods – experimental techniques

9.8.1 EVALUATION OF EXPERIMENTAL TECHNIQUES

The electrical methods involve the application of an external electric field across the specimen and the measurement of certain parameters, which are considered to be related to the diffusion coefficient of ions.

Whiting [9.71] proposed a first type of electric method for determining the chloride permeability of concrete called the **Coulomb test**, which has been adopted as AASHTO Designation T 227-83 [9.82]. A potential of 60 V d.c. is applied across a $\varnothing$ 102 × 51 mm specimen of concrete, which has been fitted between two cells as shown in Fig. 9.26 and which has been vacuum-saturated for 18 h before testing. After 6 h under test, the total charge passed through the specimen (in coulombs) is obtained by integration of the current passed through the specimen during the 6 h test

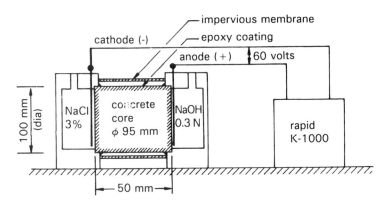

Fig. 9.26. Experimental arrangement of Coulomb test.

period. This is an indirect measurement, and its utility lies in the relation-ships between the total charge passed and the total content of chloride diffusing into a companion concrete measured by long-term immersion. It is apparently difficult to find these empirical relationships for various concrete materials. On the other hand, since total charge refers to the movement of all ions present in the pore solution (i.e. Na^+, K^+, Ca^{+2} OH^-, etc.) rather than a specific type of ion (e.g. Cl^-), it seems impossible to relate total charge to specific ion (e.g. Cl^-) diffusion.

Recently, Dhir *et al.* [9.83] modified Whiting's method by measuring the increment of chloride concentration in the anodic cell. They calculated a **potential diffusion (PD) index** on the basis of Fick's first law. However, the value of the PD index is about 100 times larger than the diffusion coefficient obtained with the conventional diffusion cell method. As they stated, the results obtained with their PD test cannot be taken directly as coefficients of diffusion.

Another type of electrical method is to measure the time needed for ions to reach the electrode embedded in a specimen. Rechberger [9.73] used an octal potentiostat to determine the diffusion coefficient of chloride ions in concrete. A 3 mm diameter steel rod was embedded in the centre of a 30 mm diameter cylindrical mortar specimen to form a 'concrete electrode'. A potential of 0.7 V was applied between a refer-ence electrode and eight concrete electrodes, which were well distributed around the reference electrode. The threshold value for chloride corro-sion was assumed as 0.2 g Cl^-/l; then the diffusion coefficient was calculated from the equation

$$D = 0.07/t \ (cm^2/s) \tag{9.18}$$

where t (s) is the time for current breakthrough due to corrosion.

The experimental arrangement of Hansen *et al.* [9.84] is shown in Fig. 9.27. The steel rod embedded in the specimen is subjected to a small impressed anodic potential of 200 mV. They use eq. (9.5), section 9.6.2, to calculate the diffusion coefficient. Neither Rechberger nor Hansen *et al.* take into account the influence of the electric field on ion diffusion in their calculation. In addition, the Cl^- threshold value for corrosion is uncertain because the amount of chloride able to depassivate the steel is a function of various parameters (pH value, amount of alkaline reserve, steel surface roughness, etc.), which are not controlled. Therefore, this type of method is inaccurate. The testing time of this method is almost similar to pure immersion experiments since, in fact, this is a kind of non-steady-state diffusion under the action of a weak electric field.

Very recently, Tang and Nilsson [9.85] established a mathematical model of ion diffusion under the action of an electric field and proposed a new type of electrical method for directly and quickly determining the

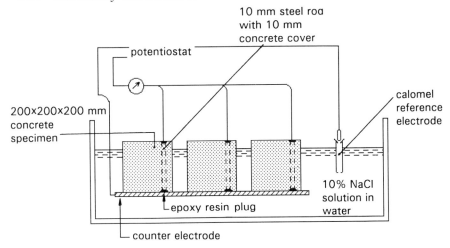

Fig. 9.27. Hansen *et al.*'s electric method [9.84].

chloride diffusion coefficient in concrete. The simple procedure involves: (1) the penetration of chlorides into a 50 mm thick specimen (cylinder or prism) by applying a potential of 30 V d.c.; (2) measurement of the chloride penetration depth by using a colorimetric method; and (3) calculation of the diffusion coefficient using the following equation, which is derived from a numerical solution of the mathematical model:

$$D = \frac{RT}{zFE} \cdot \frac{x_d - ax_d^b}{t}$$

(9.19)

where R =   gas constant (J/molK)
  T =   temperature (K)
  z =   ion valency
  F =   Faraday constant (J/Vmol)
  E =   electric field (V/m)
  x_d =   penetration depth (m)
  t =   testing time (s)
  a and b =   constants

For the chloride ion, $z = -1$. When $E = -600$ V/m and $T = 298$ K, then $a = 0.0622$ and $b = 0.589$.

Tang and Nilsson's method also belongs to the non-steady-state tests, but the diffusion process is far more accelerated by the electric field. Therefore, the testing time only needs several hours for ordinary concrete or a couple of days for high-performance concrete. As for other immersion experiments, however, the specimen should be saturated with water before testing, which usually takes a couple of days.

Table 9.4. Chloride diffusion coefficient in OPC paste

w/c	Age (day)	D (× 10^{-12} m^2/s)	Method	Researchers	
0.4	60	2.6	Steady-state	Page et al.	[9.58]
0.5	60	4.47	Steady-state	Page et al.	[9.58]
0.6	60	12.35	Steady-state	Page et al.	[9.58]
0.5	180	5.79	Steady-state	Gautefall	[9.80]
0.7	180	9.28	Steady-state	Gautefall	[9.86]
0.9	180	19.1	Steady-state	Gautefall	[9.86]
0.4	45	2.9	Steady-state	Tang and Nilsson	[9.85]
0.6	45	9.4	Steady-state	Tang and Nilsson	[9.85]
0.8	45	21	Steady-state	Tang and Nilsson	[9.85]
0.5	90	16	Non-steady-state	Gautefall	[9.87]
0.7	90	17	Non-steady-state	Gautefall	[9.87]
0.9	90	39	Non-steady-state	Gautefall	[9.87]
0.4	90	9.3	Electrical	Tang and Nilsson	[9.85]
0.6	90	17.5	Electrical	Tang and Nilsson	[9.85]
0.8	90	25.4	Electrical	Tang and Nilsson	[9.85]

9.8.2 DISCUSSION

For comparison, some experimental results from different researchers and different methods are listed in Table 9.4. It can be seen from Table 9.4 that the results from non-steady-state methods and electrical methods are consistent and higher than those obtained from the steady-state method.

Conventionally, the steady-state method is considered an accurate method [9.88]. However, its accuracy is questionable, since it is difficult to evaluate the influence of interfacial differences in concentration, which results in a lower calculated diffusion coefficient [9.85]. The results of this method are strongly influenced by some factors, such as thickness of specimen, concentration of solution, temperature, co-existing ions, etc., especially thickness which affects not only the results but also the testing time. The thicker the specimen, the longer the testing time. Jackson and Brookbanks [9.89] reported that when the thickness is about 15 mm, about 17 weeks should be spent to reach a steady state. On the other hand, it is difficult to get a thinner sample without distorting the pore structure of the specimen to be tested.

Non-steady-state methods are not only time-consuming but also labour-intensive. The main question involved in the calculation of diffusion coefficient from the testing data is how to take ion binding capacity into account. The correct description of Fick's law for diffusion accompanied by ion binding should be

$$\frac{\partial c_{\text{free}}}{\partial t} + \frac{\partial c_{\text{bound}}}{\partial t} = \frac{\partial}{\partial x} D_{\text{free}} \frac{\partial c_{\text{free}}}{\partial x} \tag{9.20}$$

or

$$\frac{\partial c_{\text{free}}}{\partial t}\left(1 + \frac{\partial c_{\text{bound}}}{\partial c_{\text{free}}}\right) = \frac{\partial}{\partial x} D_{\text{free}} \frac{\partial c_{\text{free}}}{\partial x}$$

(9.21)

which gives

$$\frac{\partial c_{\text{free}}}{\partial t} = \frac{\partial}{\partial x} D_{\text{eff}} \frac{\partial c_{\text{free}}}{\partial x}$$

(9.22)

where

$$D_{\text{eff}} = \frac{D_{\text{free}}}{\left(1 + \dfrac{\partial c_{\text{bound}}}{\partial c_{\text{free}}}\right)}$$

(9.23)

where $\partial c_{\text{bound}}/\partial c_{\text{free}}$ can be defined as ion binding capacity. As $\partial c_{\text{bound}}/\partial c_{\text{free}}$ is not necessarily a constant, the analytical solution of eq. (9.21) is not available so far. Another question is how to dissociate the free ion profile from the total ion profile obtained from the experiment. Although some researchers [9.90] have reported the method of squeezing pore solution, it is rather difficult to take a sample sufficiently large for squeezing from different depths of the specimen.

It should be noted that D_{free} and D_{eff} are different. D_{free} is traditionally determined by a steady-state test and D_{eff} is determined by non-steady-state methods, including the binding process.

Since concrete is an age-dependent material, a rapid method is necessary for both laboratory and in-situ evaluation of its properties. It is apparent that conventional steady-state and non-steady-state methods cannot meet this requirement. Electrical methods provide the possibility of rapid evaluation. Unfortunately, a Coulomb test gives wrong information regarding the desired parameter: chloride permeability. The methods proposed by Rechberger [9.73] and Hansen et al. [9.84] do not provide a rapid evaluation due to a rather low electric field applied.

9.9 References

9.1. Whiting, D. (1988) Permeability of selected concretes, ACI Special Publication 108, Eds Whiting, D. and Walitt, A., pp. 195–222.

9.2. Cook, H.K. (1951) Permeability tests of lean mass concrete, in *Proc. American Society for Testing and Materials*, Vol. 51, pp. 1156–1165.

9.3. Ludirdja, D., Berger, R.L. and Young, F.J. (1989) Simple method for measuring water permeability of concrete. *ACI Materials Journal*, Vol. 86, No. 5, pp. 433–439.

9.4. Hope, B.B. and Malhotra, M.V. (1984) The measurement of concrete permeability. *Canadian Journal of Civil Engineering*, Vol. 11, pp. 287–292.

9.5. Gjörv, O.E. (1983) *Permeabilitet som Kvalitetskrav til Betongens Bestandighet*, Report No. BML 83.603, Inst. for Bygningsmateriallære, Norway.

9.6. Sundbom, S., Olsson, K. and Johansson, L. (1987) *Betongs bestandighet*, Report No. 87953, Cement och Betong Institutet, Sweden.

9.7. Markestad, A. *An Investigation of Concrete in Regard to Permeability Problems and Factors Influencing the Results of Permeability Tests*. Report No. SFT 65A77027, Cement and Concrete Research Institute, Norway.

9.8. Powers, C., Copeland, L.E., Hayes, J.C. and Mann, H.M. (1954) Permeability of Portland cement paste. *Journal of American Concrete Institute*, Vol. 26, No. 3, pp. 285–298.

9.9. American Petroleum Institute (1952) *Recommended Practice for Determining Permeability of Porous Media*, RP 27, 3rd Edn, API, Dallas.

9.10. Collins, J.F. Jr., Derucher, K.N. and Korfiatis, G.P. (1986) Permeability of concrete mixtures. *Civil Engineering for Practising and Design Engineers*, Vol. 5, pp. 639–667.

9.11. Gräf, H. and Grube, H. (1986) Verfahren zur Prüfung der Durchlässigkeit von Mörtel und Beton gegenüber Gasen und Wasser. *Betontechnische Berichte*, Vol. 36, No. 6, pp. 222–226.

9.12. Hooten, D. and Wakeley, L.D. (1988) Influence of test conditions on water permeability of concrete in a triaxial cell, in *Materials Research Society Symposium Proceedings*, Vol. 137, Eds Roberts, L.R. and Skalny, J.P., pp. 157–164.

9.13. Ruettgers, A., Vidal, E.N. and Wing, S.P. (1935) An investigation of the permeability of mass concrete with particular reference to Boulder Dam. *Journal, ACI Proceedings*, pp. 382–415.

9.14. Dhir, R.K., Munday, J.G.L., Ho, N.Y. and Tham, K.W. (1986) Pfa in structural precast concrete: measurement of permeability. *Concrete*, December, pp. 4–8.

9.15. The Concrete Society (1985) *Permeability of Concrete and its Control*. Papers for a one day Conference, London, 12 December.

9.16. Hearn, N. and Mills, R.H. (1991) A simple permeameter for water or gas flow. *Cement and Concrete Research*, Vol. 21, No. 2.

9.17. El-Dieb, A.S. and Hooton, R.D. (1994) A high pressure triaxial cell with improved measurement sensitivity for saturated water permeability of high performance concrete. *Cement and Concrete Research*, Vol. 24.

9.18. Janssen, D.J. (1988) Laboratory permeability measurement, ACI Special Publication 108, Eds Whiting, D. and Walitt, A., pp. 145–158.

9.19. Powers, T.C. (1958) Structure and physical properties of hardened Portland cement paste. *Journal of the American Ceramic Society*, Vol. 41, No. 1.

9.20. Bager, D.H. (1983) Hardened cement paste and concrete as a living material from a point of view of pore structure, in *Proc. International Conference on Materials Science and Restoration*, Ed F.H. Wittmann.

9.21. Gjörv, O.E. and Löland, K.E. (1980) Effect of air on the hydraulic conductivity of concrete, ASTM Special Technical Publication 691, pp. 410–422.

9.22. Private communication between Prof. Hilsdorf and Prof. Mills, University of Toronto.

9.23. Jefferis, S.A. and Mangabhai, R.J. (1988) The divided flow permeameter, in

254 Laboratory test methods

Materials Research Society Symposium Proceedings, Vol. 137. Eds Roberts, L.R. and Skalny, J.P., pp. 209–214.

9.24. Roy, D.M. (1988) Relationships between permeability, porosity, diffusion and microstructure of cement pastes, mortar and concrete at different temperatures, in *Materials Research Society Symposium Proceedings*, Vol. 137. Eds Roberts, L.R. and Skalny, J.P. pp. 179–190.

9.25. Glanville, W.H. (1931) *The Permeability of Portland Cement Concrete*, Building Research, Technical Paper No. 3.

9.26. Johansson, L., Sundbom, S., and Woltze, K. (1989) *Permeabilitet, provning och inverkan på betongs beständighet*, CBI-report No. S-100 44, Cement och Betoninstituttet, Stockholm.

9.27. ISO/DIS 7031 (1983) Concrete hardened – Determination of the Depth of Penetration of Water under Pressure. Draft International Standard.

9.28. DIN 1048 (1978) Prüfverfahren für Beton, Frischbeton, Festbeton gesondert hergestellter Probekörper.

9.29. RILEM Tentative Recommendations (1979) Concrete test Methods. CPC 13.2 Test for Permeability of Porous Concrete.

9.30. ISO/DIS 7032 (1983) Concrete Hardened – Determination of Permeability. Draft International Standard.

9.31. The Concrete Society (1988) *Permeability Testing of Site Concrete: A Review of Methods and Experience*, The Concrete Society, London, Technical Report 31, 96 pp.

9.32. Tyler, I.L. and Erlin, B. (1961) A proposed simple test method for determining the permeability of concrete, *Journal of PCA*, September, pp. 2–7.

9.33. Dawei Mo, Chaozong Zhang, Zhiping Guo, Yisi Liu, Fulin AN and Qitian Mio, The application of neutron radiography to the measurement of the water-permeability of concrete, pp. 255–262, No. 79 in the list of references by Molin, C. and Rockström, J. dated 23 Nov. 1989.

9.34. Valanta, O. (1970) Durability of concrete. *Materials and Structures*, Vol. 3, No. 17, pp. 333–345.

9.35. Arnold, S.R. and Littleton, I. (1983) *Investigation into the Relationship between Aggregate Absorption and the Permeability of Concrete*, Shrivenham, Royal Military College of Science, Technical Note 12, 18 pp., May.

9.36. Bonzel, J. v. (1966) Der Einfluß des Zementes, des w/z-Wertes, des Alters und der Lagerung auf die Wasserundurchlässigkeit des Betons, *Betontechnische Berichte*, Ed. Walz, K., Beton-Verlag, pp. 144–169.

9.37. Hall, C. (1989) Water sorptivity of mortars and concretes, a review. *Magazine of Concrete Research*, Vol. 41, No. 147, pp. 51–61.

9.38. Swiss Federal Laboratories for Materials Testing and Research, (1989) SIA 162/1, Test No. 5 Water Conductivity, Guidelines for Testing.

9.39. Nycander, P. (1954) *Provning af Vattentäthet hos Betong med Prismaformade Provkroppar*, Statens Provningsanstalt, Stockholm. Meddelande 112.

9.40. Morrison, G.L., Gilliland, W.J., Bukovatz, J.E., Jayaprakash, G.P. and Seitz, R.D. *An Evapo-Transmission Method for Determining Relative Permeability of Concrete*, National Inf. Service, Springfield, Virginia, 22161.

9.41. RILEM Tentative Recommendation No. 11.2. (1974) Absorption of Water by Capillarity. *Materials and Structures*, Vol. 7, pp. 295–297.

9.42. Kollek, J.J. (1989) The determination of the permeability of concrete to oxygen by the Cembureau method – a recommendation. *Materials and Structures*, No. 22, pp. 225–230.

9.43. DIN 51 058 Bestimmung der spezifischen Gasdurchlässigkeit feuerfester Steine.

9.44. Zagar, L. (1955) Die Grundlagen zur Ermittlung der Gasdurchlässigkeit von feuerfesten Baustoffen. *Archiv für das Eisenhüttenwesen*, Vol. 26, No. 12, pp. 777–782.

9.45. Schönlin, K.F. (1989) Permeabilität als Kennwert der Dauerhaftigkeit von Beton. Dissertation TH Karlsruhe.

9.46. Kropp, J. (1983) Karbonatisierung und Transportvorgänge in Zementstein. Dissertation TH Karlsruhe.

9.47. Gräf, H. and Grube, H. (1986) Einfluß der Zusammensetzung und der Nachbehandlung des Betons auf seine Gasdurchlässigkeit. *Beton*, Vol. 36, No. 11, pp. 426–429 and No. 12, pp. 473–476; ebenso *Betontechnische Berichte* 1986–88, Beton-Verlag, Düsseldorf 1989, pp. 79–100.

9.48. Lawrence, C.D. (1984) Transport of oxygen through concrete. British Ceramic Society Meeting, Chemistry and chemically-related properties of cement, London, 12-13 April.

9.49. DIN 52 615 (1987) Wärmeschutztechnische Prüfungen; Bestimmung der Wasserdampfdurchlässigkeit von Bau- und Dämmstoffen.

9.50. Spanka, G. and Grube, H. (1991) Concrete tightness against organic liquids. International Symposium Bochum, Concrete Polymer Composites, 12-14 March, pp. 219-26.

9.51. Schwiete, H.E. and Ludwig, U. (1966) Über die Bestimmung der offenen Porosität im Zementstein *Tonind. Zeitung*, 90 pp. 562 ff.

9.52. Gräf, H. and Grube, H. (1986) Verfahren zur Prüfung der Durchlässigkeit von Mörtel und Beton gegenüber Gasen und Wasser. *Beton*, Vol. 36, No. 5, pp. 184–187 and No. 6, pp. 222–226; ebenso *Betontechnische Berichte*, 1986–88, Beton-Verlag, Düsseldorf 1989, pp. 35–36.

9.53. Hurling, H. (1984) Oxygen permeability of concrete. RILEM-Seminar on the durability of concrete structures under normal outdoor exposure. CPC 14, Hannover 26-29 March.

9.54. Currie, J.A. (1960) Gaseous diffusion in porous media, Part 1: A non-steady-state method. *British Journal of Applied Physics*, Vol. 11, pp. 314 ff.

9.55. Gräf, H. (1988) Über die Porosität und die Durchlässigkeit von Zementstein, Mörtel und Beton und ihren Einfluß auf Gebrauchseigenschaften von Beton. Diss. GH Essen.

9.56. Gordon, A.R. (1945) The diaphragm cell method of measuring diffusion. *Ann. N.Y. Acad. Sci.*, Vol. 46, p. 285.

9.57. Ushiyama, H. and Goto, S. (1974) Diffusion of various ions in hardened Portland cement paste, in *Proc. 6th Intl. Congress on the Chemistry of Cement*, Moscow, Vol. II-1, pp. 331–337.

9.58. Page, C.L., Short, N.R. and Tarras, A.El. (1981) Diffusion of chloride ions in hardened cement pastes. *Cement and Concrete Research*, Vol. 11, No. 3, pp. 395–406.

9.59. Roy, D.M., Kumar, A. and Rhodes, J.P. (1986) Diffusion of chloride and cesium ions in Portland cement pastes and mortars containing blast furnace slag and fly ash, in *Use of Fly Ash, Silica Fume, Slag and Natural*

Pozzolans in Concrete, Proc. 2nd Intl. Conference, Madrid, ACI SP-91, pp. 1423–1444.

9.60. Hansson, C.M., Strunge, H., Markussen, J.B. and Frølund, T. (1985) The effect of cement type on the diffusion of chloride, Nordic Concrete Research, Publication No. 4, pp. 70–80.

9.61. Collepardi, M., Marcialis, A. and Turriziani, R. (1970) The kinetics of penetration of chloride ions into the concrete. *Il Cimento*, Vol. 4, pp. 157–164.

9.62. Crank, J. (1975) *The Mathematics of Diffusion*, 2nd edn, Clarendon Press, Oxford, p. 21

9.63. Sørensen, H. and Frederiksen, J.M. (1990) Testing and modelling of chloride penetration into concrete, Nordic Concrete Research, Research project, Trondheim.

9.64. Poulsen, E. (1990) The chloride diffusion characteristics of concrete: approximative determination by linear regression analysis, Nordic Concrete Research, Publication No. 9, pp. 124–133.

9.65. AASHTO (1980) Designation T 259-80, Standard method of test for resistance of concrete to chloride ion penetration, Amer. Assoc. of State Highway and Transportation Officials, Washington D.C.

9.66. Hachemi, A.A. Murat, M. and Cubaud, J.C. (1976) *Revue des Matériaux de Construction*, No. 700, pp. 149–155.

9.67. Gouda, V.K. and Monfore, G.E. (1965) *Journal PCA, Research and Development Laboratories*, Vol. 7, pp. 24–36.

9.68. Grimaldi, G. and Languehard, J.C. (1986) *Bulletin de Liaison des Laboratoires des Ponts et Chaussées*, May–June, pp. 79–84.

9.69. Bennett, J.E. and Schue, T.J. (1990) *Corrosion 90*, NACE, Paper No. 316.

9.70. Whiting, D. (1981) *Public Roads*, Vol. 45, pp. 101–112.

9.71. Whiting, D. (1981) *Rapid Determination of the Chloride Permeability of Concrete*, Report No. FHWA/RD-81/119, August, NTIS DB No. 82140724.

9.72. Detwiler, R.J., Kjellsen, K.O. and Gjorv, O.E. (1991) *ACI Materials Journal*, Vol. 88, pp. 19–24.

9.73. Rechberger, P. (1985) Electrochemical determination of whole chloride diffusion coefficients. *Zement-Kalk-Gips*, Vol. 38, pp. 679–684.

9.74. ASTM C1202-91 Standard Test Method for Electrical Indication of Concrete's Ability to Resist Chloride Ion Penetration.

9.75. Bablor, J.A. and Ibarz, J. (1958) *Química General Moderna*, Manual Marin Ed. Barcelona, Spain, pp. 483–486.

9.76. Bockris, J.O'M. and Reddy, A.K.N. (1974) *Modern Electrochemistry*, Plenum Press, New York.

9.77. Glasstone, S. (1947) *Textbook of Physical Chemistry*, Van Nostrand, New York.

9.78. Bard, A.J. and Faulkner, L.R. (1980) *Electrochemical Methods. Fundamentals and Applications*, John Wiley & Sons.

9.79. Costa, J.M. (1981) *Fundamentos de Electrónica. Cinética Electroquímica y sus aplicaciones*, Alhambra Universidad Ed., Spain.

9.80. Newman, J.S. (1991) *Electrochemical Systems*, Prentice Hall, Englewood Cliffs, New Jersey.

9.81. Andrade, C. and Sanjuan, M.A. Submitted to *Advances in Cement Research*.

9.82. AASHTO Designation T 277-83, (1983) Standard method of test for rapid determination of the chloride permeability of concrete, Amer. Assoc. of State Highway and Transportation Officials, Washington D.C.

9.83. Dhir, R.K., Jones, M.R., Ahmed, H.E.H. and Seneviratne, A.M.G. (1990) Rapid estimation of chloride diffusion coefficient in concrete. *Magazine of Concrete Research*, Vol. 42, No. 152, pp. 177–185.

9.84. Hansen, T.C., Jensen, H. and Johannesson, T. (1986) Chloride diffusion and corrosion initiation of steel reinforcement in fly-ash concretes. *Cement and Concrete Research*, Vol. 16, No. 5, pp. 782–784.

9.85. Tang, L. and Nilsson, L.-O. (1992) Rapid determination of the chloride diffusivity in concrete by applying an electric field, presented at ACI Convention, Boston, March (1991). *ACI Materials Journal*, January/February 1992.

9.86. Gautefall, O. (1986) Effect of condensed silica fume on the diffusion of chlorides through hardened cement paste, in *Use of Fly Ash, Silica Fume, Slag and Natural Pozzolans in Concrete*, Proc. 2nd International Conference, Madrid, ACI SP-91, pp. 991–997.

9.87. Gautefall, O. and Havdahl, J. (1989) Effect of condensed silica fume on the mechanism of chloride diffusion into hardened cement paste, in *Use of Fly Ash, Silica Fume, Slag and Natural Pozzolans in Concrete*, Proc. 3rd Intl. Conference, Trondheim, ACI SP-114, Ed. V.M. Malhotra, Vol. II, pp. 849–860.

9.88. Buenfeld, N.R. and Newman, J.B. (1987) Examination of three methods for studying ion diffusion in cement pastes, mortars and concrete. *Materials and Structures*, Vol. 20, pp. 3–10.

9.89. Jackson, P.C. and Brookbanks, P. (1989) Chloride diffusion in concretes having different degrees of curing and made using Portland cements and blended cements containing Portland cement, pulverized-fuel ash and ground granulated blastfurnace slag, in *Use of Fly Ash, Silica Fume, Slag and Natural Pozzolans in Concrete*, Proc. 3rd Intl. Conference, Trondheim, ACI SP-114, Ed. V.M. Malhotra, Supplementary Papers, pp. 641–655.

9.90. Decter, M.H., Short, N.R., Page, C.L. and Higgins, D.D. (1989) Chloride ion penetration into blended cement pastes and concrete, in *Use of Fly Ash, Silica Fume, Slag and Natural Pozzolans in Concrete*, Proc. 3rd Intl. Conference, Trondheim, ACI SP-114, Ed. V.M. Malhotra, Vol. II, pp. 1399–1411.

10 On-site test methods

K. Paulmann and Christer Molin

10.1 Capillary suction and water penetration

10.1.1 INTRODUCTION

Capillary suction is the transport of liquids in porous solids due to the action of surface tension. For water in concrete, the capillary pressure p_w exerted upon the water surface by a capillary of radius r is

$$p_w = \frac{1.5}{r}$$

with p_w in bar, r in µm. Additional forces like gravity or external pressure acting upon the liquid have to be compared with the capillary pressure. The test methods described below use pressure heads between 0.01 bar (10 cm water head) and 10 bar, which are equal to the capillary pressure in pores with radii of 150 µm and 150 nm, respectively. So, for concrete without cracks it can be said that permeation is the predominant transport mechanism for pressures of several bar, whereas with pressure heads of some 10 cm only capillary suction takes place. In this case, the exact height of pressure should have a minor or no influence on the absorption rate. However, no experimental proof of this assumption is presented.

Gravity has to be taken into account only in very large pores or cracks (or for very large suction heights, which do not usually occur in on-site measurements).

Whereas in laboratory tests of capillary suction a one-dimensional flow can easily be established by means of prismatical or cylindrical specimens, in on-site tests absorption is usually three-dimensional, with equipotential surfaces and flow lines being geometrically complicated. So, no simple analytical relation can be given between three-dimensional absorption and hydraulic parameters defined for one-dimensional flow.

Performance Criteria for Concrete Durability, edited by J. Kropp and H.K. Hilsdorf. Published in 1995 by E & FN Spon, London. ISBN 0 419 19880 6.

All on-site test methods for capillary suction of concrete found in the literature are variations of two different methods developed by Levitt and Figg, respectively. The one measures the water absorption through the surface, the other the absorption of a layer of greater thickness underneath the surface, made accessible by a drilled hole.

In all cases, the measured quantity is the cumulative absorbed water volume or the absorption rate as a function of time or at certain points in time, measured at inflow either directly (volumetrically) or indirectly (e.g. via the pressure decay in a closed volume).

Some of the test methods described below are reviewed in [10.1] and, more recently, in [10.2].

10.1.2 ISAT AND MODIFICATIONS

The original **initial surface absorption test** (ISAT) was developed by Levitt [10.3–10.5] and is described in British Standard 1881 [10.6]. The method uses a cap sealed to the concrete surface and filled with water. A pressure head of 200 mm is set up by means of a water reservoir. At certain intervals after the first contact of the surface with water, a momentary value of suction rate is measured. In detail (see Figs 10.1 and 10.2):

• The area in contact with water should be not less than 5000 mm² (i.e. diameter ≥ 80 mm for a circular cap).
• The cap can be made from any suitable material like metal or plastic. Clear plastic allows for observation of complete filling and removal of entrapped air.

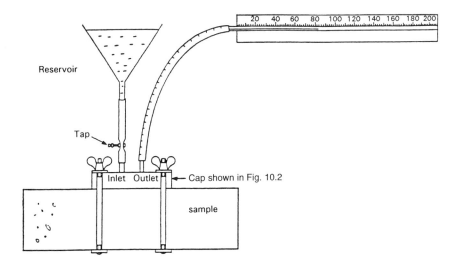

Fig. 10.1. ISAT setup [10.6].

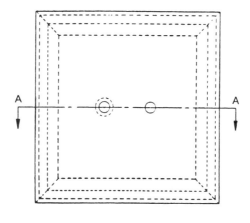

Outlet

Inlet

Square section rubber

Section A-A

Fig. 10.2. Typical cap suitable for clamping onto smooth horizontal surface.

- For sealing, the cap may have a flange with a rubber gasket or a knife edge which is embedded in moulding clay or petroleum jelly. A flange must be clamped to the surface (resulting in damage), but embedding it in clay or jelly may offer enough adhesion to support the cap even on vertical surfaces.
- A sensitive measurement of suction rate is accomplished by means of a narrow (diameter 0.8–2 mm) scaled glass capillary connected to the cap and placed horizontally at equal height with the water level of the reservoir. For the measurement, the connection between reservoir and cap is closed, so that all further absorption causes a retraction of the water meniscus in the capillary. The speed of retraction is measured and converted to ml/cm²s, which means that the size of the suction area is taken into account.
- Readings are taken 10, 30, 60 and 120 min after the start of the test.

The following modifications of ISAT have been proposed:

- Hall [10.7] points out that, owing to the three-dimensional absorption, the absorption rate per unit source area depends on the source dimension itself, and the cumulative absorption increases more rapidly with time t than $t^{0.5}$ (i.e. the absorption rate decays slower than $t^{-0.5}$). He proposes to standardize the source dimensions (preferably using a circular source to ensure axial symmetry) and to determine the slope of the cumulative absorption against $t^{0.5}$ rather than making single point measurements of absorption rate.

- For eliminating the lateral spreading of water from the source, Hall [10.2] suggests isolating a cylindrical prism of material under the source by core drilling or to use a guard ring around the source.
- A guard ring apparatus is described by Steinert [10.8]. For the investigation of cracks, he used two concentric water chambers (circular, diameter 50 mm, and ring-shaped, outer diameter 87 mm) with equal pressure, thus achieving one-dimensional flow under the inner chamber. The water is pressurized to 0.5–10 bar by means of compressed air. Measurement of absorbed water volume is done in the inner chamber only. For sealing, liquid neoprene rubber is applied to the concrete surface.
- Montgomery and Adams [10.9, 10.10] use a metal base ring bonded to the concrete surface and a water chamber fastened to it by means of screws and an O-ring. Water pressure is 1.4 bar, applied and held constant by the motion of a piston and measured with a pressure gauge. The absorbed water volume is measured by reading the motion of the piston with a micrometer screw gauge (Fig. 10.3).
- A commercially available apparatus, the Germanns Waterpermeability Test, consists of a metal housing sealed to the concrete surface by means of a compressed gasket. Water pressure between 0 and 4 bar is established by turning a nut on the housing. Pressure is kept constant and water volume is measured by means of a piston and a micrometer screw, similar to [10.9, 10.10] (Fig. 10.4).
- For measurements on masonry, a predecessor of ISAT has been widely used for some time. Mariotti and Mamillan [10.11], Kirtschig and Kasten [10.12] and others use rectangular caps with different dimensions. The applied water head is 15 cm; the water reservoir is a vertical, scaled glass tube, which also serves for measurement of cumulative absorption. Owing to the greater diameter of this tube, the resolution is poor compared with the original ISAT.

10.1.3 FIGG METHOD AND MODIFICATIONS

Figg [10.13] drilled a hole of diameter 5.5 mm and depth 30 mm into the concrete perpendicular to its surface. After thorough cleaning, the hole is plugged at part depth by polyether foam and then sealed with a catalysed silicone rubber. A hypodermic needle with two concentric tubes is pierced through the rubber after hardening, and a water head of 10 cm is applied.

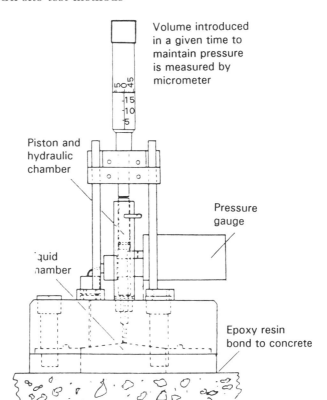

Volume introduced in a given time to maintain pressure is measured by micrometer

Piston and hydraulic chamber

Pressure gauge

Liquid chamber

Epoxy resin bond to concrete

Fig. 10.3. Apparatus by Montgomery and Adams [10.9].

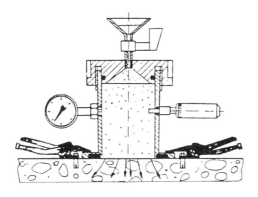

Fig. 10.4. The Germanns Waterpermeability Test.

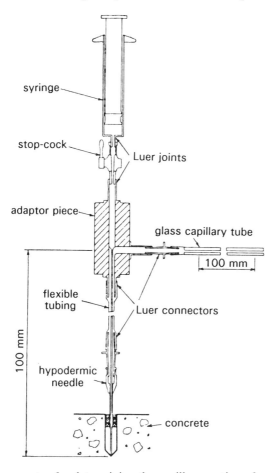

Fig. 10.5. Figg apparatus for determining the capillary suction of concrete [10.13].

The absorbed water volume is measured by motion of a meniscus in a horizontal capillary similar to ISAT (Figs 10.5, 10.6). (With a hand vacuum pump and a pressure gauge, air permeability can be measured in the same borehole.)

Improvements proposed by Figg [10.14] concern the automatic measurement of motion of the water meniscus by means of an infrared light barrier and the substitution of the hole plug cast in situ by a preformed one. Experiments with a compression-type plug (rubber annulus compressed by a screw) were not satisfactory. The following modifications of Figg's method have been suggested:

● Several authors, e.g. Pihlajavaara and Paroll [10.15], use a hole with greater diameter (e.g. 10 mm).

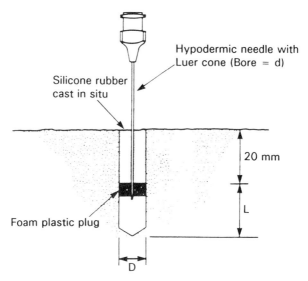

D = 5.5 mm for earlier work; 10 mm later
L = 18 mm for earlier work; 20 mm later
d = 0.51 for air permeability-test and
    1.09 mm for water-permeability test
    (cannula internal diameter = 0.50 mm)

Fig. 10.6. Figg apparatus, air- and watertight seal with concrete [10.14].

- Richards [10.16] uses a larger hole (diameter 16 mm) and two parallel tubes side by side instead of the two concentric tubes.
- Sabia [10.17] inserts a hollow metal plug with a tubular compression gasket into a hole of 10 mm diameter. A water pressure of initially 5 bar is applied and its time decay recorded for about 20 min (Fig. 10.7).
- Dhir *et al.* [10.18] find the repeatability of test results rather poor (coefficient of variation up to 40%) and attribute this to the possible entrapment of air in the tiny flow passages in the whole system.
  Accordingly they use a transparent, gasketed cap with inlet tubes of diameter 2 mm clamped over the test hole instead of the sealing inside the hole and the hypodermic needle. Moreover, the diameter and depth of the hole are enlarged to 13 mm and 50 mm, respectively. With these modifications, the coefficient of variation of test results is lowered to 8% (covercrete absorption test: CAT, see Fig. 10.8).
- Tanahashi *et al.* [10.19] developed a method applicable to walls accessible from both sides. A hole of diameter 3.5 cm is drilled throughout the wall, a hollow steel bolt inserted, and steel discs (outer diameter 15 cm) sandwiched with rubber sealings are clamped to both sides of the wall. Water is applied to the surface of the hole via the hollow bolt (Fig. 10.9).

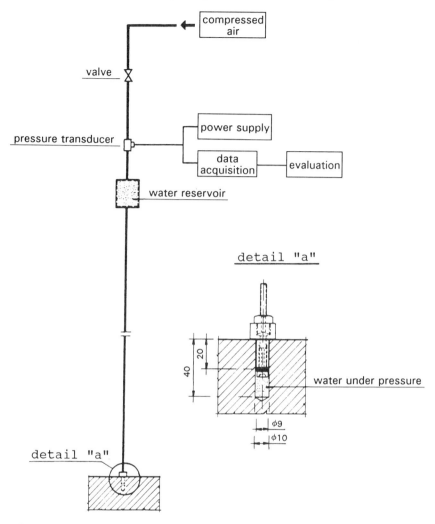

Fig. 10.7. Modification of Figg apparatus by Sabia [10.17].

Recently, a field permeability test has been developed by Meletiou and Bloomquist [10.20] at the University of Florida. As shown in Fig. 10.10, a cylindrical probe is inserted into a drilled hole, diameter 23 mm, depth 65 mm, which seals a central chamber with the help of two rubber rings. Then, water is introduced into this chamber at a pressure of 10–35 bar. After a steady-state flow is achieved, the flow rate of the water is recorded and the coefficient of the water permeability is calculated according to Darcy's law. For preconditioning of the concrete under investigation the pressurized water should soak the concrete in the vicinity of the borehole for at least 2 h before readings of the flow rate are made.

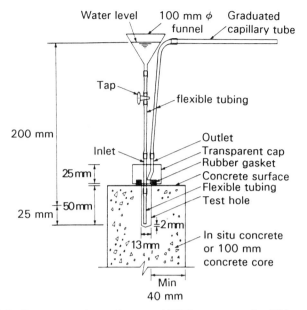

Fig. 10.8. Covercrete absorption test (CAT) apparatus by Dhir *et al.* [10.18].

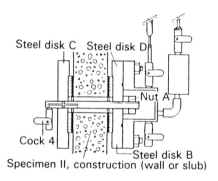

Fig. 10.9. Apparatus by Tanahashi *et al.* [10.19].

Results of different authors are not comparable due to differences in geometry. Often, suction rate is not related to the size of the suction area.

10.1.4 STANDARDIZED METHODS; COMMERCIALLY AVAILABLE APPARATUS

Standardized methods:

- ISAT in British Standard 1881: Part 5: 1970 [10.6]

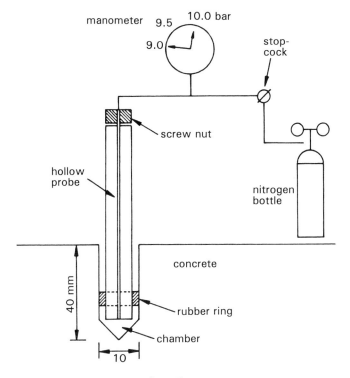

Fig. 10.10. Field permeability test [10.20].

Commercially available apparatus:

- Germanns Waterpermeability Test (modification of ISAT). Germann Instruments A/S, Copenhagen, Denmark
- Improved Figg apparatus according to [10.14], combined for air permeability and water suction. James Instruments of Chicago, Illinois (announced in [10.14] as available in the future).

10.1.5 REPRODUCIBILITY

For laboratory measurements of ISAT on dried specimens (8 batches of a concrete with $w/c = 0.55$, 10 specimens each), Dhir *et al.* [10.21] report coefficients of variation between 1.9% and 7.5% independent of the test time (10–120 min after start). Sealing a knife-edge cap with modelling clay gave a greater coefficient of variation (9.5%) than using a gasketed cap (5.5%) [10.18].

Unpublished ISAT measurements on site by Bunte and Paulmann

yielded coefficients of variation between 16% and 51%, with a tendency to decreasing variation at increasing absorption rate.

Figg [10.14] reports coefficients of variation of 13% for preformed plugs and 21% for plugs cast in situ, but for each type only three measurements were made.

Dhir et al. [10.18] found coefficients of variation up to 40% for the original Figg method and a reduction to about 8% for their improved covercrete absorption test (laboratory measurements on dried specimens, 8 batches of concrete with $w/c = 0.55$, 10 specimens each).

10.1.6 EXPERIMENTAL PARAMETERS INFLUENCING ABSORPTION

A severe problem of on-site tests of capillary suction is the difficulty of achieving a controlled and uniform initial moisture distribution. Most authors agree that particularly ISAT (and related methods) is sensitive to the moisture content of the surface layer. In consequence, it is widely accepted that ISAT measurements should be done only after at least a 48 h period without rain. Dhir et al. [10.18] recommend 7 or even 14 days. Only Montgomery [10.10] proposes pre-soaking of the surface with water overnight before measurement. Further conditioning procedures (e.g. heating) are not reported.

Theoretically, in a homogeneous solid and for one-dimensional transport by capillary forces only, the absorption rate should decrease with time t as t^{-n} (that means that the cumulative absorbed volume increases at t^{1-n}) with $n = 0.5$. Experimentally, a range $0.3 < n < 0.75$ has been observed by Levitt [10.5]. He attributes the difference to progressive changes in the pore structure during water penetration. Hall [10.7], however, shows that more-dimensional flow leads to $n \pm 0.5$ without assuming changes in pore structure.

Montgomery [10.10] reports a linear increase of absorbed water volume with time on soaked surfaces.

Especially on site, the temperature dependence of viscosity (for water far greater than for gases and with opposite tendency, decrease of about 3% per °C near 20°C) and surface tension (decrease of about 0.2% per °C) should be taken into account. The absorption rate varies with temperature as $(\sigma/)^{0.5}$ which increases by about 13% from 10°C to 20°C [10.22]. Hall [10.7] therefore recommends normalization of absorption data by applying an appropriate correction factor.

10.1.7 COMMENTS

The number of publications dealing with on-site capillary suction or water permeability is small compared to air permeability. This may be due to the following:

- In concretes that are not fully hydrated, water alters the permeability to be measured, depending on the duration of the test.
- A measurement cannot be repeated since the original moisture content is changed.
- The viscosity of water is about 100 times greater than that of gases, the volume flows being accordingly smaller.
- With liquids, only overpressure can be used.

Measurements using a liquid other than water have not been found.

10.2 Gas permeability

10.2.1 INTRODUCTION

In the literature the great importance of permeability of concrete structures is pointed out. It also appears that it is the final quality of the concrete that is decisive for the potential lifetime.

The curing of the concrete is very often considerably more favourable in laboratories than on the building site. To get a representative value of the permeability the test must be done on site and preferably by a non-destructive test. For gas permeability, the carbon dioxide and oxygen in the air are above all of interest for durability. Therefore it might be suitable to use air, or a gas that is similar to air.

It is very important to consider the moisture influence on permeability, which increases significantly as moisture is removed; see section 4.3.

There are only a few methods for testing gas permeability on site described in the literature. One method measures direct on the surface. This is completely non-destructive. The others require some kind of drilling. The permeability is consequently measured in the surface layer respectively the concrete cover of the reinforcement. Air is the prevalent medium. Both negative and positive pressures are used. The pressures are relatively low. The following methods are described in the literature.

10.2.2 FIGG

A hole, 10 mm in diameter and with a depth of 40 mm, is drilled in the concrete. The hole is made airtight with a plug and silicone. In order to prevent leakage through microcracks caused by the drilling, if any, priming of this part of the concrete face is recommended. With a hypodermic needle and hand vacuum pump the air pressure is lowered. The time for the increase of the pressure from −55 kPa to −50 kPa is measured. The apparatus is shown in principle in Fig. 10.11. The calculation of a permeability coefficient is not possible. The apparatus is commercially available and is sold with a portable case.

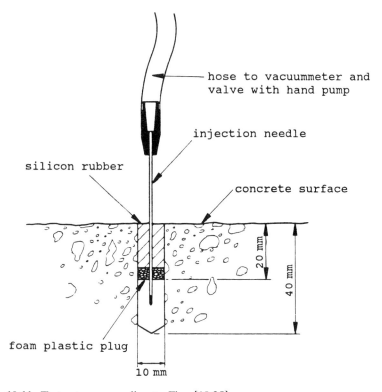

Fig. 10.11. Test set-up according to Figg [10.25].

10.2.3 KASAI

The method is a variant of Figg with a 5 mm drill hole and 10 mm plug (Fig. 10.12). The pressure difference for this method is 40 mmHg.

10.2.4 PAULMANN

A hole 11 mm in diameter and 40–45 mm depth is drilled in the concrete (Fig. 10.13). The mouth of the hole is sealed with an expanding rubber plug. A ring-shaped gas-collecting device is placed around the hole. Sealing with Vaseline is done. The hole is exposed to 2 bar overpressure by connection to a gas tube. In the first place nitrogen is used, otherwise air.

The time for a pressure increase from 0.2 mbar to 0.5 mbar is measured. The gas flow through the concrete surface in the surroundings of the hole is collected and measured by a flowmeter. A permeability measure can be estimated, which can also be used to calculate a permeability coefficient. The Paulmann method is the only true steady-state method.

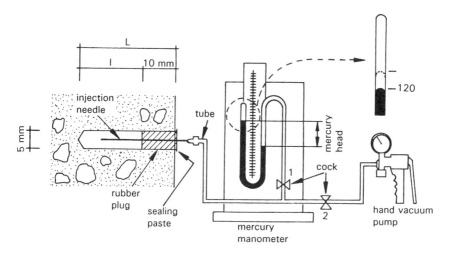

Fig. 10.12. Test set-up according to Kasai [10.28].

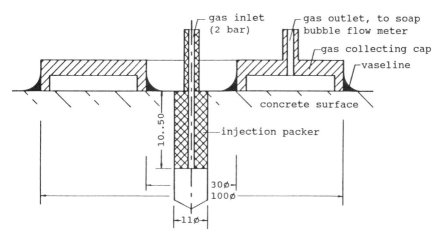

Fig. 10.13. Experimental set-up for on-site measurements of gas permeability according to Paulmann [10.29].

10.2.5 GERMANN METHOD (GGT)

A test rig is secured to the concrete surface by means of two clamping pliers fastened to the surface (Fig. 10.14). A hole is cut with a diamond tool (dry or wet) 45° to perpendicular below the 60 mm diameter test rig compression gasket with a preselected distance between the surface and the pressure sensor hole of 15–35 mm. A sensor with an airtight pressure ring is installed in the hole, checked for air tightness, and a selected CO_2

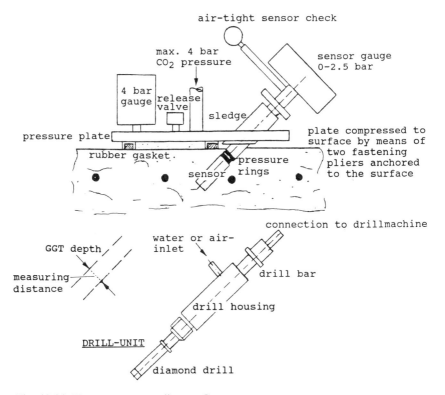

Fig. 10.14. Test set-up according to Germann.

pressure (1–4 bar) is applied to the surface. The pressure in the sensoring hole is measured over time. GGT allows the CO_2 permeability relative to the porosity to be determined. The pressure increase for a selected thickness of the cover is measured when the surface is submitted to a gas (CO_2) pressure selected between 1 to 4 bar.

10.2.6 PARROTT

A blind hole 35 mm deep and 20 mm diameter is cut into the concrete surface (Fig. 10.15). Sealing is done with stainless steel and silicone rubber plug. A pressure transducer and a digital indicator are connected to the plug. The time is taken for the air pressure in the cavity to drop over a specific range. The pressure decreases as the air permeates through the cover concrete. The geometry of air flow is measured with liquid soap applied by brushing. Therefore a permeability coefficient can be calculated. Humidity is measured in the hole for interpretation of the permeability results.

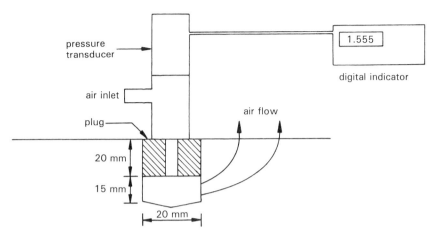

Fig. 10.15. Test set-up according to Parrott.

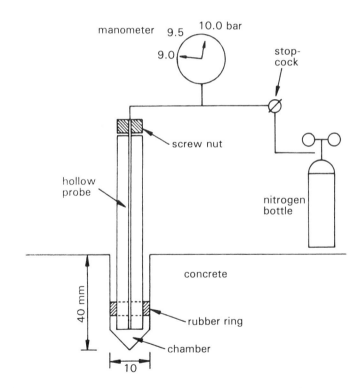

Fig. 10.16. Test set-up according to Reinhardt–Mijnsbergen [10.34].

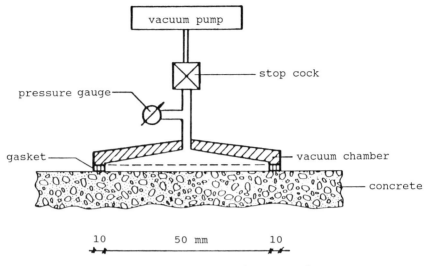

Fig. 10.17. Test set-up according to Schönlin [10.30, 10.32].

10.2.7 REINHARDT–MIJNSBERGEN

Another borehole method was introduced by Reinhardt and Mijnsbergen [10.34] using nitrogen as test gas. A sealed chamber at the tip of the bore hole is charged with nitrogen at approximately 10.5 bar (Fig. 10.16). Then the nitrogen supply is shut off and the time required for a pressure decay in the chamber from 10 down to 9.5 bar is recorded. The authors claim that, owing to the high pressure, this method is less sensitive to moisture effects than those using a low gas pressure.

10.2.8 SCHÖNLIN

See Fig. 10.17. The surface is dried with hot air. A suction cover is placed on the surface. Evacuation of air is done with a vacuum pump. The consumed time for pressure change from –50 mbar to –300 mbar is measured. For dense concrete the pressure change is read after 120 s instead. With this method a permeability index can be estimated, which is said to have good correlation with the permeability coefficient.

10.2.9 TORRENT

The main features of this method are a two-chamber vacuum cell and a regulator that balances the pressure in the inner (measuring) chamber and in the outer (guard-ring) chamber (see Fig. 10.18). Thus, unidirectional air flow into the inner chamber is achieved, eliminating any spurious

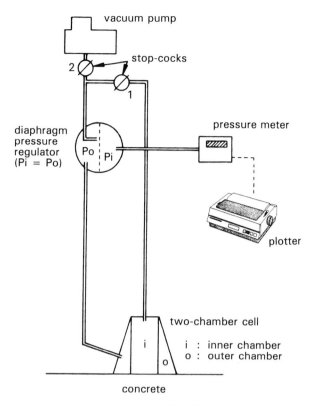

vacuum pump

stop-cocks

2

1

diaphragm
pressure
regulator
(Pi = Po)

Po | Pi

pressure meter

plotter

two-chamber cell

i

o

i : inner chamber
o : outer chamber

concrete

Fig. 10.18. Test set-up according to Torrent [10.35].

lateral air flow, which was suspected of distorting the results obtained with a single chamber. A vacuum is produced in both chambers for 1 min, when stopcock 1 is closed and the rate at which the pressure grows in the inner chamber is recorded. Owing to the better-defined gas flow geometry, this method allows a calculation of the coefficient of permeability, based on a simple theoretical model. The depth of the concrete cover affected by the test can also be estimated [10.35].

10.3 Evaluation

There are at present only two methods that are well tested and commercially available. These are Figg's method, which was developed in the early 1970s, and Schönlin's method, which has been published recently. Both methods are easy to handle on site, especially the latter, which does not need any drilling. In Schönlin's method the gas flow takes place in the surface layer of the concrete. In other methods it takes place in the covering

Table 10.1. On-site test methods

Method	Affecting parameters			Permeability coefficient K	Carbonation	Suitability for use on site	Commercially available
	Medium	RH	Measuring layer (mm)				
Figg	Air	Influence	Medium 30	No	Influence	Good	Yes
Kasai	"	"	Medium 25	No	"	Good	No ?
Paulmann	Nitrogen	"	Medium 35	Yes	"	Somewhat complicated	No
Germann	CO_2/air	"	15–35	In relation to porosity (Yes)	"	"	Yes
Schönlin	"	Considered	Surface	Yes	Much influence	Very good	Yes
Parrott	"	Measured	Medium 30	No	Influence	Good	No
Reinhardt	Nitrogen	Influence	> 40	Yes	Influence		
Torrent	Air	Influence	Surface		Influence		

layer for reinforcement. Schönlin's method is more easily disturbed by carbonation than the others. In cases when carbonation is of importance the surface must probably be ground when using Schönlin's method. The methods using a drilled hole can give an overestimated value of the tightness, as the concrete in the deepest layer normally is much denser than the layer closer to the surface. The deeper the hole is drilled the bigger this overestimation will be.

A great advantage with Schönlin's method is that it considers the great importance of relative humidity. The wet concrete surface is conditioned in a standardized way. A disadvantage is that the starting position must be a moist concrete, recently form-stripped or moist-cured. Otherwise, the concrete must be rewetted. For other methods either the relative humidity must be measured and a correction done or a conditioning method developed. A lot of development work is required before satisfactory consideration can be given to relative humidity. For concrete with RH less than about 75% the influence is considerably less than for more humid concrete. Drying down to this level should enable a realistic estimate of the permeability. The influence of the moisture is discussed in detail in section 4.2.

A summary is shown in Table 10.1.

10.4 References

10.1. The Concrete Society (1985) *Permeability testing of site concrete – a review of methods and experience*, Report of a Concrete Society working party, London 12 December, November 1985.

10.2. Hall, C. (1989) Water sorptivity of mortars and concretes: a review. *Magazine of Concrete Research*, Vol. 41, No. 147, pp. 51–61.

10.3. Levitt, M. (1970) Non-destructive testing of concrete by the initial surface absorption method, in *Proc. of the Symposium on Non-destructive testing of concrete and timber*, London, 11–12 June 1969, Inst. of Civil Engineers, pp. 23–28.

10.4. Levitt, M. (1969) An assessment of the durability of concrete by ISAT, in *Proc. of RILEM Symposium on Durability of Concrete*, Prague.

10.5. Levitt, M. (1971) The ISAT A non-destructive test for the durability of concrete. *British Journal of NDT*, July, pp. 106–112.

10.6. British Standards Institution (1970) Methods of testing hardened concrete for other than strength, BS 1881, Part 5.

10.7. Hall, C. (1981) Water movement in porous building materials iv. The initial surface absorption and the sorptivity. *Building and Environment*, Vol. 16, pp. 201 7.

10.8. Steinert, J. (1979) Zerstörungsfreie Ermittlung der Wassereindringtiefe in Kiesbeton am Bauwerk, *Forschungsbeiträge für die Baupraxis*, pp. 151–162.

10.9. Montgomery, F.R. and Adams, A. (1985) Early experience with a new concrete permeability apparatus, in *Proc. Second Int. Conf. on Structural*

Faults and Repair, London, 30 April 2 May, Edinburgh, Engineering Technics Press, pp. 359–363.

10.10. Montgomery, F.R. (1986) Concrete surface permeability, Cement and Concrete Association Research Seminar, 30 June–2 July.

10.11. Mariotti, M. and Mamillan, M. (1963) Essais in situ de perméabilité de maçonneries. *RILEM Bulletin*, 18, March, pp. 79–80.

10.12. Kirtschig, K. and Kasten, D. (1977) Zur Frage der Neuverfugung von nicht schlagregensicherem Mauerwerk. *ZI International*, Vol. 1, January, pp. 6–13.

10.13. Figg, J.W. (1973) Methods of measuring the air and water permeability of concrete *Magazine of Concrete Research*, Vol. 25, No. 85, pp. 213–219.

10.14. Figg, J.W. (1989) Concrete surface permeability: measurement and meaning. *Chemistry & Industry*, 6 November, pp. 714–719.

10.15. Philajavaara, S.E. and Paroll, H. (1975) On the correlation between permeability properties and strength of concrete. *Cement and Concrete Research*, Vol. 5, pp. 321–327.

10.16. Richards, P.W. (1982) A laboratory investigation of the water permeability and crushing strength of concrete made with and without pulverised fuel ash, as affected by early curing temperature, Slough, Cement and Concrete Association, Advanced Concrete Technology Project 82/9, 59 pp.

10.17. Sabia, D. (1989) Prova di permeabilità in sito, Politecnico di Torino, Dipartimento di Ingegneria Strutturale, Atti del Dipartimento 22, December.

10.18. Dhir, R.K., Hewlett, P.C. and Chan, Y.N. (1987) Near-surface characteristics of concrete: assessment and development of in situ test methods. *Magazine of Concrete Research*, Vol. 39, No. 141, pp. 183–195.

10.19. Tanahashi, I., Ohgishi, S., Ono, H. and Mizutani, K. Evaluation of durability for concrete in terms of watertightness by 'permeability coefficient test results', in *Concrete Durability, Katharine and Bryant Mather International Conference*, ACI.

10.20. Meletiou, C.A., Tia, M. and Bloomquist, D. (1992) Development of a field permeability test apparatus and method for concrete. *ACI Materials Journal*, Vol. 89, No. 1, January-February, pp. 83–89.

10.21. Dhir, R.K., Hewlett, P.C. and Chan, Y.N. (1984) Discussion on [10.18], reply by the authors. *Magazine of Concrete Research*, Vol. 40, No. 145, pp. 240–244.

10.22. Gummerson, R.J., Hall, C. and Hoff, W.D. (1980) Water movement in porous building materials ii. Hydraulic suction and sorptivity of brick and other masonry materials. *Building and Environment*, Vol. 15, pp. 101–108.

10.23. Cather R., Figg, J.W., Marsden, A.F. and O'Brien, T.P. (1984) Improvements to the Figg method for determining the air permeability of concrete. *Magazine of Concrete Research*, Vol. 36, No. 129.

10.24. Hansen, A.J., Ottosen, N.S. and Petersen, C.G. (1984) Gas-permeability of concrete in situ: Theory and practice.

10.25. Figg, J.W. (1973) Methods of measuring the air and water permeability of concrete. *Magazine of Concrete Research*, Vol. 25, pp. 213–219.

10.26. Johansson, L., Sundbom, S. and Woltze, K. (1984) *Permeabilitet, provning och inverkan på betongs beständighet* (Permeability, testing and durability of concrete), Cement och Betong Institutet, Stockholm.

10.27. Jönis, P.J. and Molin, C. (1988) *Mätning av betongens luftpermeabilitet* (Measuring of air-permeability of concrete), Byggnadsteknik SP Rapport: 43.

10.28. Kasai, Y., Matsui, I. and Kamohara, H. (1983) Method of rapid test for air permeability of structural concrete. *Transactions of the Japan Concrete Institute*, Vol. 5.

10.29. Paulmann, K. and Rostasy, F.S. (1989) *Praxisnahes Verfahren zur Beurteilung der Dichtigkeit oberflächennaher Betonschichten im Hinblick auf die Dauerhaftigkeit*, Institut für Baustoffe, Massivbau und Brandschutz.

10.30. Schönlin, K. (1989) *Permeabilität als Kennwert der Dauerhaftigkeit von Beton*, Institut für Massivbau und Baustofftechnologie, Karlsruhe.

10.31. Schönlin, K. *Gebrauchsanleitung für die Messung der Gasdurchlässigkeit von Betonrandzonen am Bauwerk*, Form + Test Seidner, Riedlingen.

10.32. Sundbom, S., Olsson, K. and Johansson, L. (1987) *Betongs beständighet* (Durability of concrete), Cement och Betong Institutet: Rapport nr 87053.

10.33. The Concrete Society (1985) *Permeability of concrete and its control*, Papers for a one-day conference, London, 12 December.

10.34. Reinhardt, H.W. and Mijnsbergen, J.P.G. (1989) In-situ measurement of permeability of concrete cover by overpressure, in *The Life of Structures. Physical Testing*, London, Butterworths, pp. 243–254.

10.35. Torrent, R.J. (1992) A two-chamber vacuum cell for measuring the coefficient of permeability to air of the concrete cover on site. *Materials and Structures*, Vol. 25, No. 150, July, pp 358–365.

11 Summary and conclusions

Jörg Kropp

11.1 Scope and approach

In present materials standards and building codes the durability of concrete is controlled by specifications of the quality of concrete-making materials, their proportioning for a given application as well as by requirements for the processing of the concrete, i.e. mixing, transportation, placing, compaction, and curing. Frequently, additional requirements for the compressive strength are set forth in order to ensure adequate concrete durability. These requirements are essentially based on long-term experiences with particular materials that have been used under known exposure conditions.

This empirical approach has major shortcomings: dependence on specifications for the materials used, as well as stating prescribed concrete compositions, may obstruct the introduction and use of new materials and processing methods. Thus, an empirical approach may also counteract technological progress. On the other hand, long-term experience obtained for given materials in certain areas may not be transferable to applications under differing exposure conditions or construction techniques.

It has been recognized, therefore, that rather than describing concrete quality by indirect characteristics, the degradation mechanisms occurring under various exposure conditions should be considered and minimum requirements for concrete durability should be specified for those concrete properties that control the performance of the concrete in the structure. These performance criteria could then allow the design of the required concrete durability for the planned lifetime of the structure, thereby accounting for the severity of the physical and chemical conditions prevailing at the site.

Amongst the numerous types of corrosion mechanism, severe attack of concrete structures involves in most cases the penetration of aggressive species through the concrete skin into deeper parts of a concrete section. For given types and concentrations of reactive species in the concrete the

Performance Criteria for Concrete Durability, edited by J. Kropp and H.K. Hilsdorf. Published in 1995 by E & FN Spon, London. ISBN 0 419 19880 6.

supply of the penetrating species to the reaction front will then control the corrosion rate. Therefore, the perviousness of the concrete, especially the concrete skin, controls the durability of the concrete member.

The perviousness of a concrete section can be expressed by transport coefficients for various media. However, it is necessary to distinguish between different transport mechanisms such as diffusion, permeation, and capillary suction; between different media such as ions, liquids, or gases; and even between different species such as chloride or sulphate ions. Then, for a given type of attack, a respective transport coefficient may serve as a criterion for the concrete durability, i.e. its resistance against the ingress of hazardous materials.

The approach to deriving performance criteria for concrete durability from transport coefficients may serve for different purposes at different periods during the construction process or at different ages of a structure:

- Transport coefficients may be used as a research tool in modelling corrosion rates on the basis of numerical methods.
- The potential durability of different concrete mixes may be evaluated on the basis of their transport coefficients, thus supporting the design of concrete mixes tailored for a specific application.
- In materials standards and building codes, minimum requirements for different transport characteristics may be specified. Then, transport coefficients may be determined to evaluate the compliance of a given concrete with relevant specifications or they may serve as an acceptance criterion. Depending on the purpose of the test the transport coefficients are determined in the laboratory on separately cast companion specimens, on drilled cores taken from the structure or directly on site during the erection of the structure.
- Transport coefficients may also be used in the evaluation and rehabilitation of older structures, as well as for estimating the remaining service life. In this case, it must be taken into account that the concrete under investigation has been subjected to a corrosive attack already, which may have influenced the transport coefficient under investigation.

Although not explicitly expressed in the individual chapters of this report the main emphasis of the committee's work was directed to the measurement of transport characteristics as an instrument for quality control in the design stage as well as during the construction process. However, it is preferable that the same measurement methods should also be applicable to existing structures.

11.2 The role of transport properties in corrosive actions

From the mechanisms of the individual corrosion processes it follows that theoretical correlations should exist between the transport coefficient for a given aggressive or damage-promoting agent and the corrosion rate of a particular concrete. Depending on the type of corrosive agent, the relevant transport mechanisms include:

- the diffusion of gas molecules such as CO_2, O_2, or water vapour in the gaseous phase, i.e. through empty pores of the matrix, microcracks, and interfaces;
- the diffusion of ions in the concrete pore solution, e.g. chloride or sulphate ions, as well as the diffusion of dissolved gases in the pore solution;
- the permeation of water or aqueous solutions under the influence of a hydraulic pressure head, resulting in a saturated or non-saturated capillary flow;
- the capillary suction of water or aqueous solutions occurring in empty or at least non-saturated capillaries upon contact with the liquid.

In many practical cases combinations of transport mechanisms – mixed modes – will be decisive, as shown in the following.

The permeation of gases driven by an absolute pressure head does not occur in ordinary concrete structures exposed to natural or common industrial conditions and, therefore, is not of immediate relevance for a corrosion reaction. However, it may be the case in special applications for concrete such as for airtight vessels or gas evolution in containments for active waste disposal. These rare cases will not be discussed further in this report.

The above-mentioned transport mechanisms are involved in a number of different corrosion reactions, and correspondingly, correlations should be expected between the corrosion rate and respective transport coefficients, as discussed below.

Carbonation of concrete is a diffusion-controlled process, and the diffusion coefficient for CO_2 through carbonated concrete represents the rate-controlling factor. Since the diffusion coefficient for CO_2 depends strongly on the moisture content of the concrete, which in turn is influenced by environmental conditions, carbonation is also influenced by the parameters that control the moisture state of the concrete, i.e. the take-up of water by capillary suction and the loss of water due to diffusion of water vapour in the drying process.

Sulphate attack requires the ingress of sulphate ions into the concrete. For continuously submersed concrete sections the diffusion coefficient for sulphate ions is the relevant parameter. However, sections subjected to frequent drying will absorb sulphate solutions by capillary suction. In addition to the diffusion coefficient for the sulphate ion, the absorption rate for salt solutions controls the ion ingress. For low salt concentrations the absorption of water may be taken as a characteristic value.

Chloride penetration into concrete follows the same principles as given for sulphates.

For **alkali–silica reactions**, basically all reaction species are already present in a concrete; however, the corrosion rate is enhanced by an additional ingress of moisture as well as by alkali ions. The relevant transport parameters are the capillary suction of water and diffusion coefficients for alkali ions.

Frost damage of concrete may occur if the concrete pore system is critically saturated with water. This degree of saturation can be achieved by the capillary suction of water.

Leaching and **soft water attack** will weaken the matrix by removal of soluble compounds, which are then carried out of the concrete by a convective flow together with permeating water, or by a diffusion process. The coefficient of water permeability and diffusion coefficients for the ions concerned describe the corrosion rate.

Acid attack is directed to the immediate surface of a concrete section. This attack may correlate to a transport parameter only if the resulting reaction product remains in place to form a separate layer. Then, the diffusion coefficient for the surface layer must be considered.

Corrosion of reinforcement depends on a variety of individual transport processes: depassivation of the steel surface as the prerequisite for anodic iron dissolution occurs as a consequence of carbonation or chloride ion contamination. The cathodic reaction in the corrosion cell further depends on the oxygen supply by diffusion, and a sufficient moisture for the electrolyte may be gained by capillary suction of water.

11.3 Experimentally verified correlations between transport properties and durability characteristics

Although not a relevant transport mechanism for any corrosion process considered here, the permeability of concrete to air and other gases has been widely investigated as a criterion of concrete durability. In these studies it was found that close correlations may be established under given test conditions between the gas permeability and:

- rate of carbonation;
- weight loss of concrete samples due to frost attack;

- abrasion;
- depth of chloride penetration.

In a similar manner, the capillary suction behaviour of concretes was compared with the performance of concrete, and close correlations were found with

- carbonation rate;
- weight gain upon sulphate attack;
- weight loss after frost attack;
- abrasion;
- chloride ion ingress.

Although the diffusion of molecules and ions is a very important mechanism in many corrosion reactions, these transport characteristics have not been extensively tested for a correlation with concrete durability. The diffusion of water vapour exhibited a poorer correlation to abrasion resistance than gas permeability, and the corrosion of reinforcement was controlled by oxygen diffusion only in special exposure conditions.

Mechanical wear and abrasion of concrete surfaces are corrosion mechanisms that do not depend on any transport of species through the concrete. Nevertheless, very close correlations have been reported with the gas permeability as well as with the capillary suction rate for abrasive stresses acting on the matrix of the concrete. This empirical correlation may be explained by the porosity of the paste: a very porous paste will exhibit a high gas permeability while a low strength will result in excessive abrasion.

11.3.1 ADDITIONAL INFORMATION NEEDED

Although in experimental investigations close correlations between transport properties and durability characteristics have been demonstrated, additional information on the materials used is required to quantify these relations. This is apparent in cases where selective corrosive reactions occur with individual constituents of the concrete. Then, the corrosion rate will not only depend on the supply of penetrating species but also on the reactivity as well as the concentration of the reactive phase in the concrete. As an example, the carbonation rate of concrete depends on the type of cement used, i.e. its CaO content. In a similar manner, sulphate attack depends on the availability of calcium aluminate phases in the matrix; thus the C_3A content of the cement is a major parameter. Furthermore, chloride ions penetrating into a matrix may be partially bound in chemical interactions with hydration products or by physical adsorption. The amount of chlorides thus immobilized depends on the type of cement as well as on carbonation of the matrix. Alkali–aggregate reactions in the first place depend on the vulnerability of the aggregates used, the concentration of the reactive phases and the size distribution of the aggregate particles; furthermore, the concentration of alkalis contained in the cement or additions must be considered.

For all types of chemical attack there is evidence that dispersed mineral additions such as fly ash, silica fume, ground granulated blastfurnace slag or other types of pozzolana or hydraulic compounds may not only influence the perviousness of concrete but can also interfere with the chemical reactions during hydration and in the corrosion reactions. The overall effect of these materials, therefore, may not be explained solely by transport properties.

The frost and frost/deicing salt resistance of concrete may be improved by a protective air void system in the matrix. It is unlikely that such a beneficial effect of an air void system is represented in a measured transport coefficient. Therefore, details of the air void system are additionally required to evaluate the durability of concrete subjected to frost/deicing salt attack.

The hardness of aggregates is a decisive parameter for the abrasion resistance of concrete in those cases where the abrading attack is not focused on the matrix.

11.4   Cross-correlations between transport parameters

In the published experimental investigations the permeability for gases as well as the capillary suction behaviour has been studied extensively, although gas permeability does not represent a corrosion relevant transport process. On the other hand, transport processes that are very important for corrosion mechanisms, such as the diffusion of gas molecules or ions as well as the permeation of water or solutions, have been neglected.

The obvious preference for testing gas permeability and capillary suction of concrete may result from the relative ease of the test itself and the short time required to obtain the results. Also, gas permeability tests can be used repeatedly to monitor time-dependent changes without significantly influencing the concrete microstructure. Looking at diffusion experiments with either gases or ions, sophisticated testing equipment or a test duration of up to several months are required to obtain experimental results. Furthermore, the small quantities of medium that are transported in a diffusion experiment require sensitive analysers or sensors. Also, testing the water permeability of concrete is often avoided because long test durations are necessary. These transport mechanisms may be investigated for research purposes but generally do not serve for routine testing. Therefore, it is important to establish correlations between these transport parameters and transport coefficients that can be determined easily in short-term tests.

For all transport processes through concrete a continuous network of flow paths must be available, e.g. the interconnected pore system. However, depending on the medium flowing, this interconnected pore system should either be empty of the pore solution or it should provide a continuous water film if not saturated with water. For intermediate moisture states, the flow of gases decreases with increasing moisture content; at the same time, the flow of ions and water increases. This will affect all correlations given.

Assuming that the pore system in concrete is available for all the different transport mechanisms and considering their respective physical laws, theoretical correlations could be established between the permeability K of concrete and the diffusion coefficient D for a medium in the form

$$K = \text{constant} \times D^b$$

In this relation, a value of $b = 2$ might be expected if permeation and diffusion are influenced by the pore system in the same way. However, experimental results show that different b values exist depending on the medium involved in either the permeation or the diffusion process. Nevertheless, correlations between the permeability for water and gases with the diffusion coefficients for water vapour (dry range), different gas molecules and various ions were reported and these correlations are supported by theoretical considerations. The permeability of concrete to a liquid, e.g. water, and to gases should result in the same coefficient of permeability if all physical properties of the media flowing are taken into consideration. The equivalence has been shown for coefficients of permeability higher than approximately 10^{-15} m^2. For very dense concretes, gas permeability is higher than water permeability.

Close correlations also exist between the diffusion coefficients for different gas molecules or for different ions in the pore solution. According to the kinetic theory of gases, the correlation between the diffusion coefficients for different gas molecules depends on their molecular weight, and for dry conditions this relation has been verified. The diffusion coefficients for ions in bulk water depend on their ionic mobilities and valencies. However, in the pore solution, the presence of other ions will cause additional effects. Nevertheless, strong correlations have been found between chloride and sodium ion diffusion as well as between chloride and sulphate ion diffusion.

The rate of capillary suction of water or solutions has not been closely linked to other transport parameters on a theoretical basis, but experimental results indicate that the suction rate increases with increasing gas permeability.

11.5 Recommended transport parameters and test methods

11.5.1 GENERAL

Characterizing the resistance of concrete to corrosive actions by means of transport parameters is most promising if the transport mechanisms leading to the corrosion damage are investigated directly and under the same conditions that are expected to prevail during the exposure of the concrete on the site. In other words no correlations are required because one step of the corrosion mechanism is tested directly. This approach may eventually be chosen in research projects but for various reasons will not be feasible in routine testing. The concept of quality control requires that the potential durability of a concrete can be evaluated on the basis of one or a few transport coefficients that can be determined at an early age and in a short period of time, preferably with a simple and robust test procedure. Because not all of the transport coefficients that may eventually be important can be taken into account, this approach depends on the existence of cross-correlations between the transport parameters tested and those parameters relevant for a particular type of corrosion.

The present knowledge on suitable test methods and correlations between the individual transport parameters justifies the extensive use of gas permeability measurements to characterize concrete transport properties. Although not directly relevant to degradation, this parameter offers close correlations with the diffusion coefficient for gases, the diffusion of aggressive ions in the liquid phase as well as with the permeability for water or diluted solutions. Therefore, this single parameter may characterize the perviousness of concrete in a variety of different cases, thereby covering various corrosion mechanisms.

The capillary suction of concrete has not been directly correlated with its gas permeability on a theoretical basis. However, capillary suction of concrete is considered a decisive mechanism for the uptake of water and salt solutions, thus directly affecting the resistance of concrete against frost attack, chloride contamination and eventually sulphate attack. Therefore, the capillary suction behaviour should be measured in parallel with gas permeability tests in those cases where a corresponding attack can be expected.

The permeation of water under a hydraulic pressure head may be involved in a corrosion mechanism only in rare cases. Although the water permeability of concrete then represents a direct indication of concrete perviousness for liquids, this test has become less attractive with the wider use of gas permeability measurements. This development is mainly caused by experimental difficulties due to generally higher pressures needed, the possibility of parallel microstructural changes and the long testing time required in the measurement of water permeability. Recent test methods claim to overcome these disadvantages with a special specimen conditioning and modern test equipment. A broader experience with these methods should be available before an evaluation can be made.

Non-steady-state methods that observe the penetration of a water front into non-saturated concrete involve a mixed mode of water transport, and in fact they represent a combination of a capillary suction experiment with water permeability testing. At this point, tests focusing on one single transport mechanism are preferred.

Special applications of concrete may require the knowledge of its diffusion coefficient for chloride ions, and it is questionable whether an estimate of the diffusion behaviour with the help of gas permeability tests is sufficiently precise, e.g. to calculate the projected chloride penetration in service life prediction models. Undoubtedly, conventional diffusion cell measurements in stationary transport conditions will offer a reliable and accurate method for determining the required parameters. However, the tests are difficult to perform and require weeks or even months of testing time. Similar constraints exist for immersion tests. Therefore these tests will not be applicable for routine testing. Methods for accelerated diffusion testing apply electrical fields to support the driving force, and theories have been developed to calculate either characteristic values or effective diffusion coefficients from corresponding experiments. It has been stressed that the well-known rapid chloride permeability test according to the AASHTO Designation T 227-83 does not provide sufficient information on the chloride penetration unless calibration is made for the particular concrete, for example with an immersion test. Under such conditions the test may be suitable for a fast but only qualitative evaluation of a concrete with a given constant composition.

Later developments on electrical methods and their theoretical treatment in stationary as well as non-steady-state conditions focus on the determination of the diffusion coefficient. These methods are comparatively new, and more practical experience with these techniques is desirable before an evaluation can be made. However, their potential for the characterization of concrete is recognized and further research work on these methods is recommended before they may be applied in routine testing.

11.5.2 POTENTIAL TEST METHODS

(a) Gas permeability

For the determination of the gas permeability of concrete the test method proposed by Cembureau is widely accepted, and extensive experience has been gained in many laboratories. In this method a one-directional gas flow due to an absolute pressure is applied across a cylindrical concrete sample, which is fitted into a pressure cell. The coefficient of gas permeability is calculated from the observed steady-state gas flow. The applied pressure gradient can be adjusted to the perviousness of the concrete in a certain range.

Minor modifications of this method improved the sealing of the sample on the circumference to exclude leakage in the pressure cell as well as different flowmeters.

Methods operating with vacuum versus atmospheric pressure are limited to a very small pressure gradient of less than 0.5 bar. For very dense concretes these methods are no longer applicable.

Although initially developed for on-site application, borehole methods as well as surface-fixed devices may also be used in laboratory experiments. However, they do not offer decisive advantages as compared to the Cembureau method, whereas some of these methods work on a more qualitative basis only, e.g. due to the less well-defined flow paths of the permeating gas. If corresponding methods are to be applied for on-site measurements alone or in combination with laboratory testing of companion specimens, test methods will be preferred that result in the coefficient of permeability instead of a qualitative index. Simultaneous testing of the moisture condition in the borehole will offer valuable additional information.

(b) Capillary suction

The capillary suction rate of concrete is easily tested on cast specimens or drilled cores in the laboratory by bringing one side of the sample into contact with water, thereby avoiding as much as possible the action of an additional hydraulic pressure head. This can be achieved best in the capillary rise case. For tests limited to the short-term suction rate of the concrete, i.e. the take-up of water during the first few hours only, it may not be necessary to consider the evaporation of the absorbed water through the remaining open surfaces of the specimens. Also, tests that extend much beyond 6 h involve penetration of concrete at depths that are of limited relevance to the durability of the concrete skin.

Other methods that are designed to function also for on-site measurements frequently depend on boreholes, and a small hydraulic pressure head is applied. For many of these methods a three-dimensional flow of the water occurs, and therefore the observed absorption of water may result in a qualitative absorption index. A unidirectional flow of water can be enforced by the use of guard ring techniques.

As stated earlier, tests on the diffusion of gases or ions, electrical field methods and tests on the water permeability of concrete are unsuitable for routine testing because of the long testing time required, because sophisticated test equipment is needed or because they have been proposed rather recently so that no broad support and acceptance has been established so far. Some of them may well be applicable in routine testing after more experience has been gained.

11.5.3 MOISTURE CONDITION OF THE TEST SPECIMENS

It has been pointed out that the determination of the gas permeability and the capillary absorption of concrete bears the advantage of close correlations with corrosion mechanisms and with other corrosion-related transport characteristics. Furthermore, the necessary tests are easy to conduct within short periods of time. However, a very serious problem for both measurements is the required moisture conditioning of the test specimens.

Concrete saturated with water is practically impermeable to gases because all the possible flow paths are totally blocked. On drying, the pore system will gradually empty and an increasing volume of the pore space will be made available for the flow of gases. The gas permeability strongly depends on the concrete moisture content, and it should reach a maximum for completely dry concrete when all pores are empty and are thus permeable. However, dry concrete does not exist under conditions usually encountered in service. Furthermore, drying concrete in order to liberate all free water will induce changes in the microstructure and micro-cracking. The coefficient of gas permeability observed on very dry concrete may represent a pre-damaged material under conditions that do not prevail in service.

For gas permeability measurements on concrete an intermediate moisture content must be considered that, on the one hand, is close to conditions prevailing permanently or periodically during the service of the structure and, on the other hand, will allow the characterization of a material property. A suitable moisture condition could be expected if equilibrium with a relative humidity in the range 60–80% relative humidity is obtained. Then, the coarse capillary pore system that controls the gas permeability should at least be partially emptied of water, and a gas flow is possible through the free pore space. However, further investigations are still needed to verify this assumption.

Similar considerations are valid for tests on the capillary absorption of concrete. Also, for this transport mechanism, a condition of water saturation no longer allows transport, and a completely dry condition is not desirable for the above-mentioned reasons. An intermediate moisture condition is appropriate.

In contrast to the measurement of gas permeability and capillary suction, ion diffusion in conventional cell technique, immersion tests or electrical field methods, as well as tests on the water permeability of concrete, will require specimens in the water-saturated condition. This condition is well defined and comparatively easy to achieve. In view of this advantage these methods should not be completely disregarded.

11.6   Effect of technological parameters of concrete

The review of published data has shown that the measurement of gas permeability as well as the determination of the capillary suction rate of concrete may reveal the effect of technological parameters of a particular concrete on its perviousness for a particular transport mechanism. The following parameters have been investigated:

- type of cement;
- cement content;
- water/cement ratio;
- type and grading of aggregates;
- type and duration of curing;
- age of the concrete at the time of testing; and
- use of additions.

Variations in these parameters reflected themselves in the measured transport coefficients. However, it has been impossible to compare results from different authors on a quantitative basis primarily because the measurements were conducted for different moisture states of the test specimens.

11.7   Alternative concepts

Considering alternative concepts for characterizing concrete durability it is interesting to investigate the potential of concrete compressive strength as well as the characteristics of the concrete pore structure. Both approaches may be correlated to transport properties, because compressive strength as well as transport characteristics are primarily controlled by the capillary porosity of the paste and, therefore, similar correlations with durability should be expected as for the transport parameters.

However, the standard compressive strength as a well-established quality criterion of concrete had failed to describe its durability mainly because the observed compressive strength refers to an average property over a large section of the concrete specimen, whereas the durability is controlled by the properties of the near-surface regions.

It may be assumed that the inhomogeneous structure originates from the curing effect: i.e. hydration of the cement ceases at the termination of curing while the centre sections of a concrete specimen or member continue to hydrate and thus acquire a denser structure. An approach was therefore made to consider as a characteristic value the concrete strength at the end of curing instead of a 28-day strength. For this case good correlations were observed with the development of carbonation of different concretes.

In principle, characteristics of the pore structure of concrete could also serve as criteria for the potential durability because all transport processes occur in the pore system. Although the individual transport mechanisms may be affected in different ways by these characteristics, the basic relations for modelling the transport through an idealized pore system have been established. Aside from total pore volume, pore size distribution, and threshold pore radii, the overruling parameter of the pore system is the continuity of the capillary pores. In recent research work theoretical models have been established to predict the volume fraction of the interconnected capillary pores as a function of technological parameters such as degree of hydration and water/cement ratio, and an extension of these models to concrete is ongoing. There, interpenetrating contact zones between aggregates and the matrix are considered, which form a second continuous flow path system through the concrete. These models are very promising, but much experimental work is still needed for their calibration and verification.

The experimental determination of characteristics of the pore system may serve for calibration or verification of these models or may be investigated for a direct correlation with durability characteristics. It is important to note, however, that the decisive characteristic of the pore system, i.e. the continuity, can be determined only by means of a permeation test.

11.8 Further work needed

Evaluation of the existing data has shown that in many cases a comparison of test results from different laboratories failed because different test methods and procedures were applied. The differing moisture content of the specimens for tests on gas permeability and capillary suction has been of greater significance than the selection of a particular experimental set-up for testing. It was pointed out that harmonization of the test methods and of the preconditioning of specimens is urgently needed.

Further research work is required on the moisture content of the test specimens; procedures should be defined for the necessary preconditioning of the specimens to obtain the required moisture content and its homogeneous distribution across the specimens. These procedures should allow a reproducible conditioning of the specimens within a short period of time, but without distorting the material properties or removing the effect of technological parameters, e.g. curing. Thus, preconditioning is a crucial step in the approach, and research on the necessary procedures should be given priority.

Further research work on test methods should concentrate on a critical evaluation of the large number of different test methods, and a systematic comparison should reveal their resolution (spread) for different concrete grades as well as the reproducibility of the results. Also, accelerated diffusion tests in an electrical field may need further research work for their verification.

Much experimental research is still needed on direct corrosion experiments for different types of attack. A wide span of concrete qualities should be tested, and the different concretes must be characterized by their transport coefficients. Then, a comprehensive set of quantitative correlations may be established between corrosion rates and transport coefficients of concretes. The concept of quality control will require that limiting values for the transport characteristics must be set out. These criteria must be classified for different exposure conditions for the concretes on the basis of these corrosion tests.

Appendix A: A review of methods to determine the moisture conditions in concrete

Leslie Parrott

A wide range of methods for the laboratory and in-situ measurement of moisture conditions in concrete and related materials are reviewed in terms of their purpose, principles and practicalities. The effects of sampling volume, sample geometry, pore fluid composition and moisture gradients are considered. Methods based upon destructive sampling, relative humidity, resistivity, dielectric properties, thermal properties, infrared absorption and neutron scattering are reviewed. Destructive sampling and relative humidity measurements can provide reliable data that is relevant to a wide range of concrete properties, including durability and permeation. Microwave and neutron scattering methods seem to warrant further development for moisture measurements in concrete.

A.1 Introduction

There is an increasing realization in journals and at technical meetings and conferences that moisture in concrete plays a critical role in controlling the properties of concrete and the performance of concrete structures. In particular, the durability of concrete and the corrosion of reinforcing steel are intimately linked to the moisture history of exposed concrete surfaces. Understanding of these issues is highly relevant to the development of European codes of concrete practice that can convincingly account for the effects of climate upon durability. This development is impeded by lack of information on methods of moisture measurement and their application to problems of concrete durability. On a more specific level the interpretation of permeability measurements for the control and monitoring of durability is hindered if moisture conditions in the permeated concrete are not known [A.1]. There are many other aspects of

Performance Criteria for Concrete Durability, edited by J. Kropp and H.K. Hilsdorf. Published in 1995 by E & FN Spon, London. ISBN 0 419 19880 6.

concrete performance that would benefit from knowledge of the moisture conditions in concrete [A.2–A.4]. Two examples that carry important financial implications are the dependence of the energy efficiency of buildings upon the moisture content of the shell and the deterioration of applied finishes due to moisture transfer in service. This review is intended to encourage the development of moisture measurement methods and their use for solving moisture-related problems of concrete performance.

The most appropriate definition of moisture state depends upon the particular concrete property being considered. For example, the rate of chloride or sulphate ingress from an external source will depend upon the quantity of water held in capillary pores: this provides the transport channels for ion diffusion. Conversely, the rate of carbonation depends upon the pore volume from which water has been lost: this provides the transport channels for gas (i.e. carbon dioxide) diffusion. With regard to other aspects of concrete performance such as reinforcement corrosion or cement hydration it may be better to define moisture in terms of the internal relative water vapour pressure (relative humidity).

Water in concrete is held in three main states: as free water held by capillarity, as adsorbed water held by surface forces, and as bound water held chemically. These states may be conceptually associated with mobility in three, two and zero directions. Unfortunately the binding energies for these three states overlap, and in consequence it is often difficult to quantify unambiguously the water held in each state. The free water content of saturated concrete, expressed as a volume fraction, equates with the capillary porosity and is a useful indication of concrete quality. Strictly speaking, the pore water is an aqueous solution, often saturated with respect to lime, and containing a wide range of ions [A.5, A.6].

Under most exposure conditions the moisture in concrete is not uniformly distributed and the distribution varies with exposure time. Fluctuating moisture conditions occur in the surface zone of concrete and can affect the protection it offers to underlying concrete and reinforcing steel.

Furthermore, the pore structure in which water is held varies with time due to chemical and physical changes in the products of cement hydration.

The foregoing remarks indicate that the ideal method of moisture measurement should be able to quantify the amounts of free, adsorbed and bound water regardless of the composition of the aqueous pore fluid. It should also be able to monitor variations of moisture and relative humidity non-destructively as a function of time and spatial location. Not surprisingly there is no single method that meets these requirements, and even under controlled laboratory conditions a substantial range of methods would be required for a comprehensive description of moisture conditions in a sample of concrete.

Various methods of moisture measurement are reviewed in subsequent sections of this appendix in terms of how closely they meet the ideal requirements. Methods based upon destructive sampling, relative humidity, resistivity, dielectric properties, thermal properties, infrared absorption and neutron scattering are considered. Other methods such as nuclear magnetic resonance [A.7–A.10], low-temperature calorimetry involving the freezing of pore water [A.11–A.13] and gamma-ray attenuation [A.14, A.15] can be very useful for characterizing water in cementitious materials. However, they have been excluded from the review because they require highly specialized and expensive equipment. Also, in the case of gamma-ray attenuation, extreme care is required to avoid health risks. Some publications relating to soils and other porous materials have been included in the review where the methods described seem to be potentially useful for moisture measurements in concrete.

A.2 Destructive methods

Destructive sampling of concrete by drilling or fracture requires only simple equipment and permits the use of accurate, gravimetric determination of moisture content using oven drying [A.15–A.21]. Free water can also be determined by reaction with calcium carbide [A.16, A.17]. The normal oven temperature for drying is 105°C and in consequence the measured moisture content will include all free water, all adsorbed water and a small quantity of chemically bound water [A.22].

Samples may be in the form of drill powder [A.17–A.19], fractured segments [A.20] or slices taken from cores [A.21]. Carrier [A.21] reassembled core slices so that they could be replaced in a structural element for the monitoring of time-dependent moisture gradients. Figure A.1 shows some of the data obtained by Carrier to illustrate the effect of permanent formwork (SIP formed) upon moisture distribution. Owing to the use of gaskets, liquid transport across the boundaries of each slice was prevented. Thus the measured moisture profiles may not be fully representative of the undisturbed concrete. Rothwell [A.18] reported time-dependent moisture gradients in walls using repeated drilling: some of his results showing the effects of surface coatings are illustrated in Fig. A.2. The small amount of experimental scatter in Figs A.1 and A.2 indicates that gravimetric methods give very reproducible results. Newman [A.17] demonstrated that sample moisture loss associated with drilling could be small, and a repeatability of ± 1% was possible for moisture content determinations. The obvious disadvantages are that the method is time-consuming and that drilling or cutting of a structure is not always acceptable.

Nilsson [A.2] suggested that there are advantages in measuring the relative humidity over a concrete sample; often the relative humidity needs

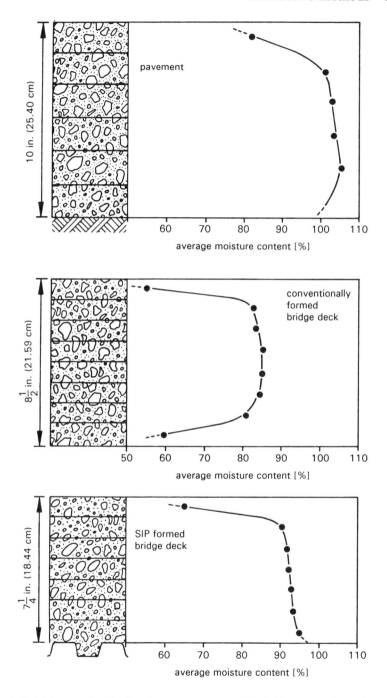

Fig. A.1. Moisture distribution in a pavement and in bridge decks [A.21.]

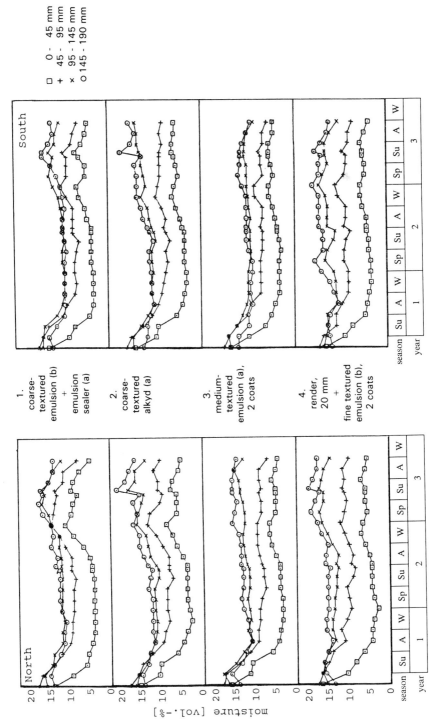

Fig. A.2. Time-dependent variations of moisture content at different depths in concrete walls [A.18].

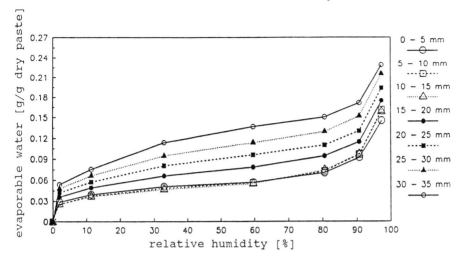

Fig. A.3. Water sorption of drill powders from different depths in a column [A.23].

to be known, the method is sensitive and reproducible, only a small sample is required and the equipment is inexpensive. Schmugge [A.15] gave a brief review of the relevant hygrometric techniques. It is often advantageous to determine the sorption isotherms for field samples so that in-situ relative humidity measurements can be more comprehensively interpreted. This can be readily accomplished using desiccators conditioned to specified relative humidities with saturated salt solutions [A.23]. Figure A.3 indicates that variations of pore structure with sample depth can cause large differences in the relationship between moisture content and relative humidity within a given concrete element.

Several investigators have shown that monitoring moisture content by gravimetric methods or by relative humidity measurements greatly aids the interpretation of permeability data [A.1]. The results of Ujike [A.24] show that in the case of air permeability it is relevant to consider the loss (i.e. empty porosity) rather than the residual quantity of evaporable water (Fig. A.4). This means that it is sometimes more important to know the saturated weight rather than the oven-dried weight.

A.3 Relative humidity in concrete

Monfore [A.25] designed a miniature relative humidity probe for measuring gradients of moisture in concrete. Monfore's work stimulated further investigations where internal relative humidity was related to fire resistance [A.26], curing [A.27, A.28], shrinkage [A.28, A.29], permeability [A.1, A.23, A.30–A.32], concrete composition [A.4] and climatic differences [A.4, A.23,

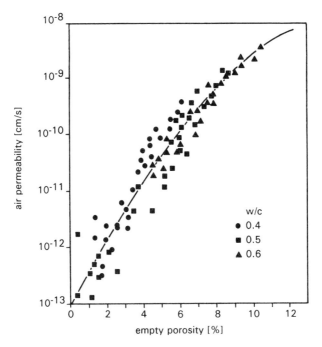

Fig. A.4. Increase of air permeability with loss of pore water; based on [A.24].

A.33, A.34]. As explained in the previous section, relative humidity results can be interpreted in terms of capillary and adsorbed water if appropriate sorption isotherms are available. Hansen [A.35] has catalogued sorption isotherms for a range of building materials and temperatures. It is also possible to judge from relative humidity results whether or not the capillary pore system is empty. This condition cannot be judged from a measurement of moisture **content** but is particularly relevant to the interpretation of gas permeability and water absorption measurements [A.1].

The sensing element of a relative humidity probe is generally a hygroscopic material in which length, resistance or capacitance changes are detected. Such probes are commercially available and are small enough to be used for measuring relative humidity gradients in concrete. Figure A.5 shows time-dependent changes in relative humidity profiles, where the 7 mm diameter measurement cavities were cast perpendicular to the axis of uniaxial drying [A.4, A.34]. It is possible to store small samples in the cavities so that changes of evaporable water content can be monitored. This allows in-situ sorption isotherms to be plotted (Fig. A.6). Where access is not available at the time of casting, a measurement cavity can be drilled into the concrete surface. The cavity is sealed over part of its length to permit relative humidity measurements at a selected depth

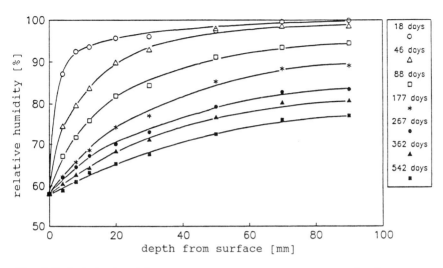

Fig. A.5. Relative humidity profiles in laboratory-dried concrete [A.4].

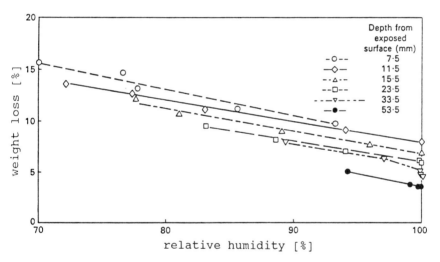

Fig. A.6. Weight loss of prisms versus relative humidity from measurements in drying concrete [A.34].

[A.2, A.23]. The use of a removable probe allows regular calibration, and this is a significant advantage for measurements in concrete, where moisture exchange is slow and measurements extend over several years. Monfore [A.25] reported that relative humidity measurements were repeatable to ± 2%, and this was consistent with the standard deviations of around 2.5% reported for measurements in nominally similar concrete specimens [A.4]. Nilsson [A.88] demonstrated that relative humidity in concrete is affected by temperature changes. The effect derives from the temperature sensitivity of the hygrometer, a small effect of temperature upon the sorption isotherm and differences in temperature between the relative humidity probe and the test material. It follows that the probe should be brought into temperature equilibrium with the test material before readings are taken. Nilsson's data suggest that relative humidity variations typically amount to 0.1–0.5% per °C.

Changes of relative humidity in porous materials are accompanied by changes of capillary tension in the pore water. Determination of this pressure can provide a sensitive measure of the energy with which water is held. The system has been used extensively in soils [A.15, A.16] but its application is limited to the higher moisture contents.

A.4 Resistivity

The resistivity of a porous material decreases with an increase of moisture content because the resistivity of the pore fluid is significantly lower than the resistivity of the solid phase. Thus moisture content changes can, in theory, be monitored simply by measuring changes in the electrical resistance of the test material. Numerous publications [A.36–A.47] show that the resistivity of concrete and mortar increases greatly with loss of pore water; the effect seems to be more pronounced at lower moisture contents, as illustrated for example in Fig. A.7 [A.39]. Possibly loss of continuity of the water in capillary channels is more important for resistivity than the volume of capillary water. The results of Alonso [A.43] show that repeated wetting and drying between saturation and 50% relative humidity causes near reversible changes of resistivity.

Unfortunately, although resistivity is very sensitive to moisture content changes it is also sensitive to a number of other factors. Resistivity increases with

- increasing cement hydration [A.36–A.38, A.40, A.42];
- decreasing water cement ratio [A.37, A.38, A.39, A.41, A.42];
- decreasing temperature [A.37, A.39, A.41, A.45, A.46] (Fig.8).

The effects of cement hydration and water/cement ratio could be accounted for via their influence upon capillary porosity and the related amount of pore fluid. According to Fig. A.8 the effect of temperature appears smaller at higher moisture contents. Resistivity is also affected by cement type [A.36, A.37] and pore fluid composition [A.37, A.38]. However, for a given concrete, resistivity measurements can conveniently provide information relating to time-dependent moisture gradients [A.40, A.41, A.47]. The volume of concrete that is sensed is comparable to the

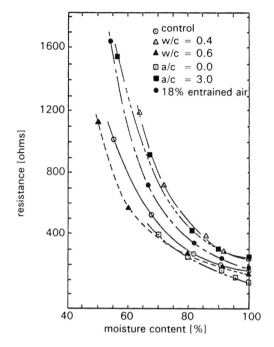

Fig. A.7. Increase of resistivity with loss of moisture [A.39].

maximum electrode spacing [A.46], and there is general agreement that control of electrode contact resistance requires special attention.

It is evident from this brief review that resistivity measurement will not give reliable estimates of moisture content without careful calibration for each test condition. If detailed knowledge of moisture gradients is required then such calibration might be justified.

A.5 Dielectric properties

The discussion in this section is limited to moisture-related measurements at microwave frequencies. Although measurements at lower frequencies are feasible [A.48] permittivity measurements at microwave frequencies (1–100 GHz) have significant advantages:

- They are specific to polar relaxation in water molecules [A.15, A.49, A.50].
- The permittivity of water is much higher than the solid phases in many porous materials [A.50], and this contributes to the sensitivity of the method.

304   Appendix A: Moisture conditions in concrete

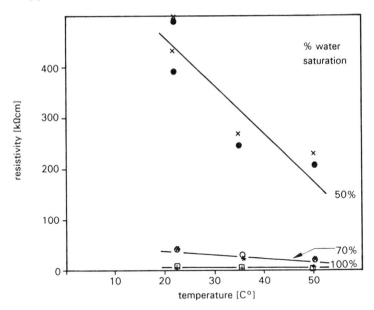

Fig. A.8. Effects of temperature and moisture on resistivity [A.45].

- Bound and adsorbed water contribute little to permittivity above about 8 GHz: they are effectively seen as solid [A.15, A.49, A.50]. Thus measurements at higher microwave frequencies reflect free, rather than total, moisture.
- Dissolved salts, which contribute to conductivity, have little effect on permittivity above frequencies of a few GHz [A.49, A.51–A.53).

Development of microwave methods for moisture measurement started around 1955 but it was not until about 1980, when inexpensive, solid-state devices for microwave generation and detection became available, that the method became more popular [A.50].

The imaginary component of the complex permittivity of water gives rise to attenuation of a penetrating microwave beam, and the real component is responsible for a reduction of wavelength and a related phase change. In a porous material both the real and imaginary components of permittivity change with moisture content. It has been found that measurements of the real and imaginary components or, more commonly, attenuation and phase change can often be combined to obtain a measure of moisture content that is independent of density [A.50, A.53–A.55]. This avoids the need for supplementary methods of density measurement [A.48].

Microwave attenuation measurements use a transmitter on one side of a sample and a receiver on the opposite side. In such a case the sampled volume will be known with reasonable certainty. Unfortunately, attenuation increases at higher frequencies, and with high moisture contents sample thickness may be restricted [A.56]. It is possible to test from one side of a sample by measurements involving microwave reflection [A.49, A.50, A.58] or the resonance of a microwave cavity [A.50, A.57, A.59, A.60]. The energy of a penetrating microwave beam diminishes exponentially with depth, and the interpretation of reflection or resonance data can be uncertain with regard to the sampled volume and moisture gradients. A twin transmitter/receiver attenuation probe was proposed [A.61] for the measurement of moisture gradients along parallel bores in a sample but no subsequent reports of a developed device were found.

Numerous publications describe the use of microwaves for measuring moisture in food [e.g. A.49, A.50, A.53–A.55] and in granular or porous materials [e.g. A.49–A.51, A.56, A.57], but there are relatively few papers relating to hardened concrete [A.52, A.60–A.64]. However, there are several basic studies of hardened cement paste that establish the feasibility of using microwaves to monitor moisture non-destructively and the effects of water/cement ratio, cement hydration and temperature [A.65–A.69]. Wittmann [A.67] reviewed much of the cement and concrete data up to 1975 and demonstrated that the decrease of evaporable water with decreasing water/cement ratio and increasing time of hydration could be monitored by attenuation of microwaves at a frequency of 10 GHz (Fig. A.9). Measurements on dried pastes indicated that attenuation due to the solid phase was relatively small. Other data also show that the chemical binding of water by cement hydration can be monitored as a function of time by changes of microwave attenuation [A.66, A.68]. Freezing [A.67] and wetting/drying [A.66, A.67] experiments also suggested that permittivity was mainly controlled by loosely held water in the larger pores (Fig. A.10).

Measurements on concrete also suggest that changes of free water content due to drying and cement hydration can be monitored using microwave attenuation [A.52, A.61–A.64], although no systematic studies of water/cement ratio, cement type, curing, compaction or temperature were found. Bhargava [A.60] described a microwave resonance method for the rapid, non-destructive estimation of moisture in concrete: access to only one concrete surface was necessary. The volume of concrete sampled was investigated using moisture-conditioned slabs stacked to give different sample thickness. The results in Figs A.11 and A.12 show that beyond a given thickness the relationship between moisture content and instrument reading (representing a change of resonant frequency) was virtually constant. The effective depth of penetration decreased from about 150 to 60 mm as the frequency was increased from 0.3 to 0.8 GHz. These data suggest that resonance measurements at higher frequencies would penetrate to a limited, but useful, depth. The effective penetration depth did not seem to depend upon moisture content. Bhargava also

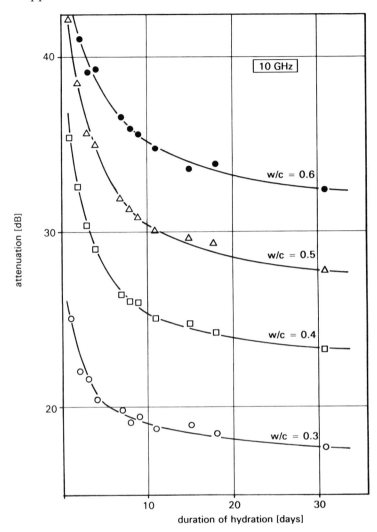

Fig. A.9. Microwave attenuation for sealed cement pastes [A.67].

reported that with frequencies of 2.4 GHz difficulties of measurement were encountered due to microwave reflection.

It seems from this review that dielectric properties have considerable potential for moisture measurement in concrete where attenuation can be measured. If phase change can also be measured then it seems likely that the reliability of moisture measurement could be improved. However, considerable research and development is required for application to concrete, and the thickness of concrete that can be tested may be a restricting factor at the desirable frequencies of a few GHz or more.

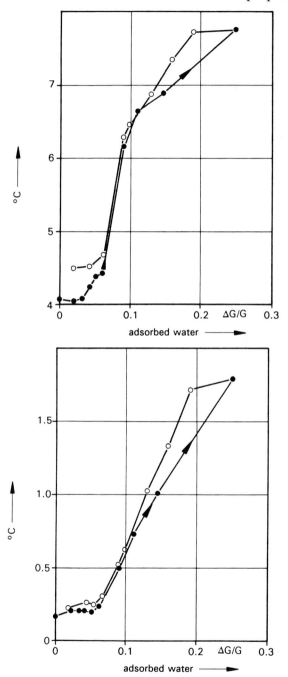

Fig. A.10. Complex permittivity of hardened cement paste (0.4 water/cement ratio) as a function of moisture content [A.67].

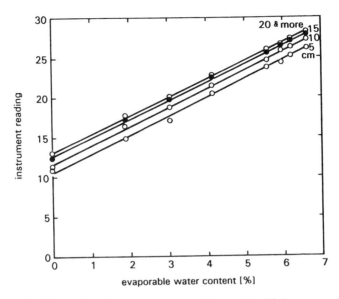

Fig. A.11. Effect of moisture content and sample thickness on change of microwave resonance frequency at 0.3 GHz [A.60].

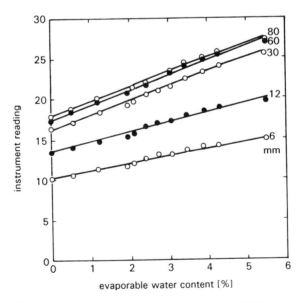

Fig. A.12. Effect of moisture content and sample thickness on change of microwave resonance frequency at 0.8 GHz [A.60].

A.6 Thermal properties

The moisture content of a porous material can quickly and conveniently be determined non-destructively from in-situ measurements of thermal conductivity or diffusivity if calibration data are available to relate thermal properties to moisture content [e.g. A.15, A.70, A.71]. Small temperature probes and heaters are implanted and the thermal response is measured as a function of time after the heater is activated. The equipment is inexpensive and is small enough to allow measurements of moisture gradients. Variations of thermal conductivity with moisture content from dry to saturated are approximately linear [A.71–A.74], so thermal conductivity reflects the total evaporable water content: it does not seem to distinguish adsorbed and free water.

A major disadvantage of using thermal properties to assess moisture content is that thermal properties also depend upon numerous additional factors, the most important being aggregate content, aggregate type, concrete density and air entrainment [A.72, A.74]. Temperature gradients imposed for the purpose of testing could cause significant changes of moisture content unless the heat flux is minimized. Furthermore the method does not yet seem to provide a very accurate measure of moisture content [A.70].

A.7 Infrared absorption

Water absorbs electromagnetic radiation at specific wavelengths in the 1–14 μm infrared portion of the spectrum. Each absorption band is characteristic of a particular vibrational mode of the water molecule. By monitoring changes of absorption at one or more wavelengths it is possible to determine the moisture content of liquids and porous solids. Only a thin layer of the sample surface is penetrated by the infrared beam, so measurements are normally undertaken in the reflection mode rather than the transmission mode; however, variations of surface texture and colour can influence results [A.15, A.48, A.77–A.80]. It is possible to position the infrared source and detector at some distance from the sample surface (50–300 mm), and this can simplify the application of the method [A.48].

Cornell [A.77] described an infrared reflection instrument operating at wavelengths of 1.94 and 1.80 μm to give moisture-sensitive and reference signals, respectively. The instrument contained sealed reference samples for calibration during use. Cornell proposed measurements on the flat-bottomed surfaces of holes drilled to different depths, as a method of investigating moisture gradients in a drying sample. Calibration curves were found to depend upon the material tested, and for brick the results were rather scattered. However, the results were not significantly affected

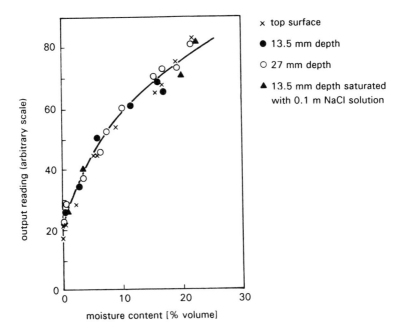

Fig. A.13. Infrared instrument readings against moisture content for 1:1:6 cement:lime:sand mortar samples cut from different depths [A.77].

by dissolved salt in the pore fluid (Fig. A.13). McFarlane [A.78] reviewed the use of infrared absorption and the choice of operating wavelengths for moisture measurements in food. Reflection methods were normally used because beam penetration into food was usually no more than 1 mm. Infrared light-emitting diodes have recently been designed specifically for the water absorption wavelengths; this, coupled with the use of fibre optic light guides, has led to simpler, lighter and more compact moisture-sensing equipment [A.48]. The application of this new equipment to moisture measurements in concrete seems limited by the low penetration depth of an infrared beam. Even if Cornell's drilling method is used, the high concentration of large pieces of aggregate at the wall of the hole (i.e. at the sample surface) would cause a diminished and variable instrument response.

Infrared radiation may be detected over a relatively large area of a structure using a video camera fitted with an infrared detector and a scanner. It is then possible to obtain a real-time image of surface temperature variations [A.79, A.80]. Wavelength bands detected are relatively broad (2–5.6 μm or 8–14 μm) and are not specific to particular vibrational modes of the water molecule. However, moisture conditions affect the heat flow through the imaged surface and corresponding surface temperature variations can be detected [A.79, A.80]. Alternatively the reflected portion of an applied infrared beam can be detected [A.79]. The interpretation of surface temperature variations in terms of moisture content is complex, but by conducting measurements at night under calm-air conditions temperature differences for the dry–wet range may amount to 3°C [A.80]; this is significantly larger than the 0.2°C resolution of current equipment [A.79].

Spatial resolution is adjustable simply by varying the distance between the sample surface and the infrared scanner. Schmugge [A.15] reported that infrared thermography has been used to remotely sense the variations of moisture in soil surfaces via the daily range of surface temperatures.

No precise relationship between moisture content and surface temperature seems possible but the infrared thermography method can assist with a broad assessment of moisture over a large area and the selection of locations for measurements using more precise techniques. The method of using an applied beam and measuring the infrared reflection in the 2–5.6 μm wavelength range has been reported by Schickert [A.79]. He showed that moisture variations at the surface of a masonry wall could be effectively imaged. The moist conditions at the base of the wall, penetration of moisture along the joints and the localized effect of an injected moisture barrier were readily seen (Fig. A.14).

A.8 Neutron scattering

When fast neutrons are emitted from a radioactive source such as americium 241: beryllium, they will penetrate concrete and collide with the nuclei of atoms composing the concrete. The dominant interaction involves elastic scattering and a reduction of kinetic energy. The velocity reduction is greatest for collisions with nuclei that have a mass comparable to that of the neutron. After a series of collisions the slow neutrons can be monitored using a detector that incorporates a slow neutron absorber. Hydrogen in water molecules is the dominant source of light nuclei that causes the production of slow neutrons in concrete. Thus, in the absence of organic material and other sources of hydrogen, the slow neutron count is primarily a measure of the total water content. The method does not distinguish chemically bound, adsorbed and free water. The common elements in cement and concrete such as silicon, calcium, aluminium and oxygen cause little absorption of slow neutrons but chlorine (e.g. in the form of sodium chloride) can reduce the slow neutron count and the indicated moisture content [A.16].

Commercial equipment often includes a source and detector for gamma radiation so that complementary measurements of density can be obtained [A.15, A.16, A.48]. The health concern associated with radioactive

Fig. A.14. Infrared reflection image of moisture (dark tone) on the surface of a masonry wall [A.79].

methods has been given careful consideration in the design of commercial equipment, but it is still necessary for equipment to be licensed, to use trained staff and to proceed with extreme care: the neutron scattering methods do not seem appropriate for routine use on a typical construction site.

The normal geometric configurations include a surface-mounted detector with a choice of surface or borehole location of the fast neutron source. Alternatively the detector and source may both be located in a borehole probe. Operation in transmission modes generally seems to be more accurate than operation in the backscatter mode. Claims for accuracy range from 0.1 to 1% of water by volume [A.16, A.48, A.81], but this level of performance is not always achieved [A.82, A.83]. Burn [A.84] examined possible materials for calibration, but samples of the test material at known moisture contents provide the most reliable calibration data [A.85].

Huet [A.81] demonstrated that moisture gradients in soils could be measured over a borehole depth of about 150 cm using a neutron moisture probe, but resolution was limited because the effective sampling volume extends over 10–60 cm [A.16]. Careful design of the neutron source and detector geometry can improve resolution [A.16], but it has been demonstrated that gamma-ray absorption is a more appropriate method for measuring steep moisture gradients [A.14, A.15, A.86].

It is possible to distinguish free and bound water using neutron scattering equipment that is fitted with high-resolution detectors [A.87]. This was a specialized laboratory method requiring sophisticated data analysis. However, it was concluded that the standard method of oven-drying at 105°C caused a substantial loss of chemically bound water from hydrated cement. This conclusion underlines an inherent difficulty in defining the moisture content of cementitious materials and providing reliable calibration data.

A.9 Discussion

None of the methods described in sections 2–8 of this appendix seems to provide a reliable and convenient means for monitoring free water profiles in concrete, and many of the methods are expensive, insensitive or unsuitable for field measurements. The possibility of measuring moisture gradients without drilling boreholes seems remote. The general characteristics of the reviewed methods are summarized in Table A.1. The assessments relate only to the reviewed publications and, because of their brief and general nature, they are somewhat subjective.

The method of drilling or cutting to obtain a sample for gravimetric analysis of moisture is more appropriate to research and development than to investigative field studies. Labour costs are high, and repeated drilling is often unacceptable for a structure in regular use.

Relative humidity measurements in a borehole are simple and convenient, but it is desirable to use the drill powder for establishing a relationship between relative humidity and evaporable moisture content.

Resistivity measurements are particularly relevant to assessments of reinforcement corrosion, but they will provide only an approximate indication of moisture content. Resistivity is influenced by many factors, especially pore fluid composition, but useful information on time-dependent changes in moisture profiles can be obtained from an array of embedded electrodes.

Dielectric methods seem particularly appropriate for measurement of free water in concrete, but there are few publications where recent technological advances have been exploited. The opposing frequency requirements of adequate sample penetration and insensitivity to dissolved salts have yet to be resolved. There is a lack of inexpensive, microwave moisture meters, and this has probably discouraged investigation of their potential for concrete.

Table A.1. Comparison of methods for moisture measurement in concrete

Method	Geometry	Approximate sampled Volume/depth	Accuracy[1]	Specific to water?	Water measured			Moisture gradients	Cost	
					Free	Absorbed	Bound		Equip	Labour
Drill or cut and standardized drying	Borehole	~ 1 to 5 cm^3	High	No	✓	✓	✓	✓	Low	High
Relative humidity	Borehole	~ 2 to 10 cm^3	Moderate	Yes	✓	✓	–	✓	low	Moderate
Resistivity	Surface	~ 100 cm	Low	No	✓	✓	–	–	Low	Low
	Embedded electrodes	Proportional to electrode spacing	Low	No	✓	✓	–	✓	Low	Moderate
Dielectric properties	Surface	~5–15 cm	Moderate	Yes	✓	–	–	–	High^2	Low
	Prismatic (attenuation)	Prism thickness	High	Yes	✓	–	–	✓	High^2	Low
Thermal properties	Embedded probe	~ 1 cm or more	Low	No	✓	✓	–	✓	Low	Low
Infrared properties	Surface probe	~ 1mm or less	High	Yes	✓	–	–	–	High	Low
	Surface thermography	~ 1 mm or less	Low	No	✓	✓	–	–	High	Low
Neutron scattering	Surface	~ 100 cm	Moderate	No	– Total water –			–	High	Low
	Borehole	~ 100 cm	Moderate	No	– Total water –			✓	High	Moderate

1 Assuming calibration data is available for test material
2 Could be reduced if commercial equipment was available

Thermal methods are particularly relevant where knowledge of thermal conductivity and thermal diffusivity is required, but they will provide only a very approximate estimate of moisture content.

Infrared absorption results can be difficult to interpret because of the variety of influencing factors and the limited sampling depth. There appear to be severe problems of representative sampling in the case of concrete, especially where moisture gradients are present. Under favourable conditions infrared thermography can provide 'whole field' imaging of surface moisture. This could be valuable for selecting locations for more detailed measurements, using other methods.

Neutron-scattering methods can detect gradients of total water content, although existing equipment does not seem ideal where gradients are steep. The influence of bound water upon the results necessitates careful calibration or the use of a parallel method that responds only to free or evaporable moisture. Considerations of safety, licensing and training would probably limit any widespread use of radioactive methods in the construction industry.

The introduction to this review indicated a specific interest in the durability and permeability of exposed surface layers of concrete. The relevance of moisture parameters to the appropriate deterioration and transport processes will now be considered. Concrete is particularly vulnerable when the pore system is close to saturation: frost attack, disruptive alkali–silica expansion, chloride ion ingress and sulphate ion ingress are then feasible. In addition, corrosion of reinforcement is likely if its normally stable oxide layer has been disrupted by carbonation or the presence of chlorides.

The critical moisture state for these processes cannot be adequately described simply in terms of a high moisture content. In the case of frost attack the volume of empty pores or water loss is equally or more important. The possibilities of disruptive alkali–silica expansion and reinforcement corrosion are more closely related to the local relative humidity.

The transport channels for chloride and sulphate ion ingress are related to both the amount and the tortuosity of the capillary liquid phase. On the other hand the rate of carbonation is maximized when the capillary pores are just emptied, and there is minimal resistance to the ingress of carbon dioxide but sufficient water for reaction. Empty porosity and relative humidity are then the two relevant moisture-related parameters.

Gas permeability and the rate of water absorption are two physical properties that are commonly expected to relate to durability. Gas permeability, like carbonation, is sensitive to the emptying of capillary pores, and this can be most easily assessed in terms of the empty porosity when the local relative humidity is 60–70% [A.1]. The rate of water absorption is difficult to interpret unless a similar condition is reached [A.1]. In summary it is often more important to know the moisture loss relative to the saturated condition than it is to know moisture content. One fortunate consequence is that the definition of a dry state and the separation of free, adsorbed and bound water become less significant issues in the definition of moisture conditions. Furthermore, it is evident that assessment of durability and permeation properties is significantly helped if relative humidity data are available.

A.10 Concluding remarks

Moisture measurement methods for concrete are severely limited in performance at present, and only two methods can be recommended with confidence: (1) drilling or cutting followed by gravimetric analysis using standardized drying, and (2) relative humidity measurements. These two methods are particularly valuable when used in combination and relationship between relative humidity and water content or water loss is established. Data generated from the use of these methods are relevant to a wide range of concrete properties, including permeation and durability.

Further investigation of microwave and neutron-scattering methods for moisture measurement in concrete seems warranted. In particular, microwave methods seem to offer a direct measure of the free water held in capillary pores. Infrared thermography can provide a broad picture of surface moisture under certain, favourable conditions and thus indicate locations suitable for more detailed measurements using other methods. It is possible that developments in remote sensing with microwave will also lead to information on near-surface moisture conditions.

A.11 References

A.1. Parrott, L.J. *Influence of environmental parameters upon permeability: a review*, State of art report of RILEM Technical Committee 116-PCD.

A.2. Nilsson, L.O. (1980) *Hygroscopic moisture in concrete – drying, measurements and related material properties*, Lund Institute of Technology, Report TVBM-1003, pp. 162.

A.3. Verbeck, G.J. and Helmuth, R.H. (1968) Structure and physical properties of cement paste, in *Proc. 5th International Symposium on Chemistry of Cement*, Tokyo, Vol. 3, pp. 1–32.

A.4. Parrott, L.J. (1990) *Factors influencing relative humidity in concrete*, British Cement Association Internal Report PP/525, June, pp 18. To be published.

A.5. Taylor, H.F.W. (1987) A method for predicting alkali ion concentrations in cement pore solutions. *Advances in Cement Research*, Vol. 1, No. 1, October, pp. 5–17.

A.6. Reardon, E.J. and Dewaele, P. (1990) Chemical model for the carbonation of a grout/water slurry. *Journal of the American Ceramic Society*, Vol. 73, No. 6, pp. 1681–1690.

A.7. Seligmann, P. (1968) Nuclear magnetic resonance studies of water in hardened cement paste. *Journal of PCA Research and Development Laboratory*, Vol. 10, No. 1, pp. 52–65.

A.8. Bahajnar, G., Blinc, R., *et al.* (1977) On the use of pulse NMR techniques for the study of cement hydration. *Cement and Concrete Research*, Vol. 7, No. 4, pp. 385–394.

A.9. Schreiner, L.J., MacTavish, J.C. *et al.* (1985) NMR line shape-spin-lattice relaxation correlation study of Portland cement hydration. *Journal of the American Ceramic Society*, Vol. 68, No. 1, pp. 10–16.

A.10. Gummerson, R., Hall, C. *et al.* (1979) Unsaturated water flow within porous materials observed by NMR imaging. *Nature*, Vol. 281, 6 September, pp. 56–57.

A.11. Bager, D. and Sellevold, E. (1986) Ice formation in hardened cement paste, Part I – Room temperature cured pastes with variable moisture contents. *Cement and Concrete Research*, Vol. 16, No. 5, pp. 709–720.

A.12. Bager, D. and Sellevold, E. (1986) Ice formation in hardened cement paste, Part II – Drying and resaturation on room temperature cured pastes. *Cement and Concrete Research*, Vol. 16, No. 6, pp. 835–844.

A.13. Beddoe, R.E. and Setzer, M.J. (1990) Phase transformations of water in hardened cement paste: a low-temperature DSC investigation. *Cement and Concrete Research*, Vol. 20, No. 2, pp. 236–242.

A.14. Nielsen, A.F. (1983) Gamma-ray attenuation used on free water intake tests, in *Proc. Autoclaved Aerated Concrete, Moisture and Properties*, Elsevier, pp. 43–53.

A.15. Schmugge, T.J., Jackson, T.J. and McKim H.L. (1980) Survey of methods for soil moisture determination. *Water Resources Research*, Vol. 16, No. 6, pp. 961–979.

A.16. OECD Road Research Group (1973) *Water in Roads: Methods for determining soil moisture content and pore water tension*, OECD Report Paris, pp. 61.

A.17. Newman, A.J. (1975) *Improvement of the drilling method for the determination of moisture content in building materials*, Building Research Establishment Report CP 22/75, pp. 7.

A.18. Rothwell, G.W. (1983) The effects of external coatings on moisture contents of autoclaved aerated concrete walls, in *Proc. Autoclaved Aerated Concretes, Moisture and Properties*, Elsevier, pp. 101–116.

A.19. Southern, J.R. (1983) Moisture in solid blockwork walls at Glenrothes, in *Proc. Autoclaved Aerated Concrete, Moisture and Properties*, Elsevier, pp. 313–322.

A.20. Terrill, J.M., Richardson, M. and Selby, A.R. (1986) Non-linear moisture profiles and shrinkage in concrete members. *Magazine of Concrete Research*, Vol. 38, No. 137, pp. 220–225.

A.21. Carrier, R.E., Pu, D.C. and Cady, P.D. (1972) *Moisture distribution in concrete bridge decks and pavements*, ACI Special Publication SP47, pp. 169–189.

A.22. Taylor, H.W.F. (1990) *Cement Chemistry*, Academic Press.

A.23. Parrott, L.J. (1990) Assessing carbonation in concrete structures, in *Durability of Building Materials and Components*, Proc. 5th International Conference, Brighton, E & FN Spon, London.

A.24. Ujike, I. and Nagataki, S. (1988) A study on the quantitative evaluation of air permeability of concrete. *Proc. Japanese Society of Civil Engineers*, Vol. 9, No. 396, pp. 79–87.

A.25. Monfore, G.E. (1963) A small probe-type gauge for measuring relative humidity. *Journal of the PCA Research and Development Labs*, Vol. 5, No. 2, pp. 41–47.

A.26. Abrams, M.S. and Monfore, G.E. (1965) Application of a small probe-type relative humidity gauge to research on fire resistance of concrete. *Journal of the PCA Research and Development Labs*, Vol. 7, No. 3, pp. 2–12.

A.27. Carrier, R. and Cady, P. (1970) Evaluating effectiveness of concrete curing compounds. *Journal of Materials*, Vol. 5, No. 2, pp. 294–302.

A.28. Hanson, J.A. (1968) Effects of curing and drying environments on splitting tensile strength of concrete. *Journal of the American Concrete Institute*, Vol. 65, No. 7, pp. 535–543.

A.29. Arumugasaamy, P. and Swamy, R. (1978) Moisture movements in reinforced concrete columns. *Il Cemento*, Vol. 75, No. 3, pp. 121–128.

A.30. Parrott, L.J. and Chen Zhang Hong (1991) Some factors influencing air permeation measurements in cover concrete. *Materials and Structures*, Vol. 24, No. 144, pp. 403–409.

A.31. Chen Zhang Hong and Parrott, L.J. (1989) *Air permeability of cover concrete and the effect of curing*, British Cement Association Report C/5, October, pp. 28.

A.32. Jonis, J. and Molin, C. (1988) *Measuring of air permeability of concrete*, Swedish National Testing Institute, Report 1988: 4.

A.33. Abrams, M.S. and Orals, D.L. (1965) *Concrete drying methods and their effect on fire resistance*, American Society for Testing and Materials, ASTM Report STP 585, pp. 52–73.

A.34. Parrott, L.J. (1988) Moisture profiles in drying concrete. *Advances in Cement Research*, Vol. 1, No. 3, pp. 164–170.

A.35. Hansen, K. (1987) Sorption isotherms – a catalogue and a data base, in *Proc. Symposium on Building Physics in the Nordic Countries,* Lund, August, pp. 369–373.

A.36. Hammond, E. and Robson, T.D. (1955) Comparison of electrical properties of various cements and concretes. *The Engineer*, Vol. 199, January, pp. 78–80 and 114–115.

A.37. Monfore, G.E. (1968) The electrical resistivity of concrete. *Journal of the PCA Research and Development Labs*, May, pp. 35–48.

A.38. Gjorv, O.E., Vennesland, U. and El-Busaidy, U. (1977) Electrical resistivity of concrete in the oceans, Offshore Technology Conference, Houston, May, pp. 581–588.

A.39. Woelfe, G.A. and Lauer, K. (1979) The electrical resistivity of concrete with emphasis on the use of electrical resistance for measuring moisture content. *Cement, Concrete and Aggregates*, Vol. 1, No. 2, pp. 64–67.

A.40. Sriravindrarajah, R. and Swamy, R.N. (1982) Development of a conductivity probe to monitor setting time and moisture movement in concrete. *Cement, Concrete and Aggregates*, Vol. 4, No. 2, pp. 73–80.

A.41. Hope, B.B., Ip, A.K. and Manning, D.G. (1985) Corrosion and electrical impedance in concrete. *Cement and Concrete Research*, Vol. 15, No. 3, pp. 525–534.

A.42. Hughes, B.P., Soleit, A.K.O. and Brierley, R.W. (1985) New technique for determining the electrical resistivity of concrete. *Magazine of Concrete Research*, Vol. 37, No. 133, pp. 243–248.

A.43. Alonso, C., Andrade, C. and Gonzalez, J. (1988) Relation between resistivity and corrosion rate of reinforcement in carbonated mortar made with several cement types. *Cement and Concrete Research*, Vol. 18, No. 5, pp. 687–698.

A.44. Nagataki, S. and Ujike, I. (1988) Effect of heating condition on air permeability of concrete at elevated temperature. *Trans. Japanese Concrete Institute*, Vol. 10, pp. 147–154.

A.45. Millard, S.G. (1989) Effects of temperature and moisture upon concrete permeability and resistivity measurements. Workshop on in situ measurement of concrete permeability, University of Loughborough, 12 December, pp. 9.

A.46. Millard, S.G., Harrison, J.A. and Edwards, A.J. (1989) Measurement of the electrical resistivity of reinforced concrete structures for the assessment of corrosion risk. *British Journal of Non-Destructive Testing*, Vol. 31, No. 11, pp. 617–621.

A.47. Tritthard, J. (1990) Pore solution composition and other factors influencing the corrosion risk of reinforcement in concrete, in *Corrosion of Reinforcement in Concrete*, Proceedings of SCI Conference, Wishaw, Page, C.L., Treadaway, K.W.J. and Bamforth, P.B. (Eds) Elsevier, pp. 96–106.

A.48. Carr-Brion, K. (1986) *Moisture Sensors in Process Control*, Elsevier, pp. 122.

A.49. Thompson, F. (1989) Moisture measurements using microwaves. *Measurement and Control*, Vol. 22, No. 7, pp. 210–215.

A.50. Kraszowski, A. (1980) Microwave aquametry – a review. *Journal of Microwave Power*, Vol. 15, No. 4, pp. 209–220.

A.51. Hallikainen, M.T., Ulaby, F.T. *et al.* (1985) Microwave dielectric behaviour of wet soil. *IEEE Trans. Geoscience and Remote Sensing*, Vol. GE-23, No. 1, pp. 25–34.

A.52. Watson, A. (1970) Measurement and control of moisture content by microwave absorption. *Build International*, March, pp. 47–50.

A.53. Meyer, W. and Schilz, W. (1980) A microwave method for density independent determination of the moisture control of solids. *J. Physics, D: Applied Physics*, Vol. 13, pp. 1823–1830.

A.54. Kent, H. and Meyer, W. (1982) A density-independent microwave moisture meter for heterogeneous foodstuffs. *Journal of Food Engineering*, No. 1, pp. 31–42.

A.55. Kent, M. (1989) Applications of two-variable microwave techniques to composition analysis problems. *Trans. Inst. Meas. Control*, Vol. 11, No. 2, pp. 58–62.

A.56. Cutmore, N., Abernathy, D. and Evans, T. (1898) Microwave technique for the on-line determination of moisture in coal. *J. Microwave Power and Electromagnetic Energy*, Vol. 24, No. 2, pp. 79–90.

A.57. Kondrat'yev, Y.F., Fedorov, V.V. and Slobodyanik, V.H. (1988) Dielectric constant of microwave frequencies and measurement of the moisture content of sea-bottom sediments. *Oceanology*, Vol. 28, No. 1, pp. 115–117.

A.58. Gabriel, C. and Grant, E.H. (1989) Dielectric sensors for industrial microwave measurement and control. *Microwellen & HF Magazine*, Vol. 15, No. 8, pp. 643–645.

A.59. de Jongh, P.F. (1989) Moisture measurements with microwaves. *Microwellen & HF Magazine*, Vol. 15, No. 8, pp. 648–649.

A.60. Bhargava, J. and Lundberg, K. (1972) Determination of moisture content of concrete by microwave-resonance method. *Materials and Structures*, Vol. 5, No. 27, pp. 165–168.

A.61. Building Research Station, DSIR (1962) The non-destructive measurement of water content by microwave absorption: further developments. *RILEM Bulletin*, No. 15, June, pp. 85–87.

A.62. Figg, J. (1972) Determining the water content of concrete panels by using a microwave moisture meter. *Magazine of Concrete Research*, Vol. 24, No. 79, pp. 93–96.

A.63. Palletta, F. and Ricca, A.M. (1986) Concrete moisture evaluation by microwaves. *Alta Frequenza*, Vol. 55, No. 4, pp. 255–264.

A.64. Mathey, R. (1962) Application du rayonnement électromagnétic a l'etude de l'eau dans les matériaux. *RILEM Bulletin*, No. 15, June, pp. 138–143.

A.65. de Loor, G.P. (1961) Some dielectric measurements on Portland cement paste. Physics Lab. Report of National Defence Research Organization, TNO, Report 1961-14.

A.66. Schlude, F. and Wittmann, F.H. (1972) Measurement of the moisture content of porous materials by microwave absorption, in *Proc. Conf. High Frequency Dielectric Measurement*, National Physical Laboratory, March, pp. 140–146.

A.67. Wittmann, F.H. and Schlude, F. (1975) Microwave absorption of hardened cement paste. *Cement and Concrete Research*, Vol. 5, No. 1, pp. 63–71.

A.68. Raboul, J.P. (1978) The hydraulic reaction of tricalcium silicate observed by microwave dielectric measurements. *Revue de Physique Appliquée*, Vol. 13, No. 8, pp. 383–386.

A.69. Gorur, K., Smit, M.K. and Wittmann, F.H. (1982) Microwave study of hydrating cement paste at early age *Cement and Concrete Research*. Vol. 12, No. 4, pp. 447–454.

A.70. Woodbury, K.A. and Thomas, W.C. (1985) Measurement of moisture concentration in Fibrous insulation using a microprocessor-based thermistor probe, in *Proc. Symp. Moisture and Humidity*, Washington DC, April, pp. 467–74.

A.71. Quenard, D., Derrien, F. and Cope, R. (1984) Migration d'humidite, in *Durability of Building Materials and Components*, Proc. 3rd International Conference, Espoo, Vol. 1, pp. 462–479.

A.72. Loudon, A.G. (1983) The effect of moisture content on thermal conductivity, in *Proc. Conf. Autoclaved Aerated Concrete*, Lausanne, pp. 131–142.

A.73. Tinker, J.A. (1988) Modelling the thermal properties of lightweight aggregate masonry materials containing moisture, in *Brick and Block Masonry*, Vol. 1, Elsevier, pp. 42–51.

A.74. Hums, D. (1983) Relation between humidity and heat conductivity in aerated concrete, in *Proc. Conf. Autoclaved Aerated Concrete*, Lausanne, pp. 143–151.

A.75. Tokuda, H. (1984) Thermal properties of concrete. *Concrete Journal*, Vol. 22, No. 3, March, pp. 29–37.

A.76. Neville, A.M. (1971) Thermal properties. Chapter 16 of *Hardened Concrete:*

Physical and Mechanical Aspects, American Concrete Institute Monograph No. 6, pp. 189–206.

A.77. Cornell, J.B. and Coote, A.T. (1972) The application of an infrared absorption technique to the measurement of moisture content of building materials. *Journal of Applied Chem. Biotechnology*, Vol. 22, pp. 455–463.

A.78. McFarlane, I. (1984) Moisture measurement in food by infrared absorption, in *Proc. Transducer Tempcon Conferences*, Harrogate, pp. 169–179.

A.79. Schikert, G. (1985) Infrared thermography as a possible tool to detect damaged areas in buildings, in *Durability of Building Materials*, Vol. 3, pp. 87–99.

A.80. Jenkins, D.R., Mathey, R.G. and Knab, L.I. (1981) *Moisture detection in roofing by non-destructive means – A state of art survey*, National Bureau of Standards Report TN 1146, July, pp. 82.

A.81. Huet, J. (1962) Détermination de la teneur en eau des sols en place par la méthode neutronique. *RILEM Bulletin*, No. 15, June, pp. 37–46.

A.82. Clark, A.J. (1983) In situ density measurements of pavements using a nuclear gauge. Institute of Concrete Technology Convention, Nottingham, April (C&CA Report PP/353), pp. 13.

A.83. Hundt, J. (1977) Heat and moisture conduction in concrete. *Deutscher Ausschuß für Stahlbeton*, No. 280, pp. 23–41.

A.84. Burn, K.N. (1962) A neutron meter for measuring moisture in soils. *RILEM Bulletin*, No. 15, June, pp. 91–97.

A.85. Waters, E.H. (1965) Measurement of moisture in concrete and masonry with special reference to neutron scattering techniques. *Nuclear Structural Engineering*, Vol. 2, pp. 494–500.

A.86. Kumaran, M.K, and Bomberg, M. (1985) A gamma-spectrometer for determination of density distribution and moisture in building materials, in *Proc. Symposium Moisture and Humidity*, Washington D.C., April, pp. 484–490.

A.87. Harris, D.H.C., Windsor, C.G. and Lawrence, C.D. (1974) Free and bound water in cement paste. *Magazine of Concrete Research*, Vol. 26, No. 87, pp. 65–72.

A.88. Nilsson, L.O. (1987) Temperature affects in relative humidity measurements on concrete – some preliminary studies, in *Proc. Symposium on Building Physics*, Lund, August, pp. 456–462.

Appendix B: Conversion[a] table for permeability[b] units (1)

Table B.1. Conversion[a] table for permeability[b] units (1)

	darcy	millidarcy	m/s	m²	Meinzers	ft/day
darcy	1	1000	9.68×10^{-6}	9.87×10^{-13}	20.50	2.75
millidarcy	10^3	1	9.68×10^{-9}	9.87×10^{-16}	2.05×10^{-2}	2.75×10^{-3}
m/s	1.03×10^1	1.03×10^4	1	1.03×10^{-11}	2.12×10^2	2.84×10^1
m²	1.01×10^{12}	$1/01 \times 10^{15}$	9.71×10^6	1	20.7×10^{12}	2.78×10^{12}
Meinzers	4.88×10^{-2}	48.78	4.72×10^{-7}	4.83×10^{-14}	1	1.34×10^{-1}
ft/day	3.64×10^{-1}	3.64×10^2	3.52×10^{-6}	3.60×10^{-13}	7.46	1

[a] To convert from units in column at left to units at top multiply by indicated factor
[b] Conversions given are appropriate for cases of saturated, steady-state flow. For units associated with diffusion processes, or with relative, empirical test procedure, no *direct* conversions are available.

Performance Criteria for Concrete Durability, edited by J. Kropp and H.K. Hilsdorf. Published in 1995 by E & FN Spon, London. ISBN 0 419 19880 6.

Index

RILEM

RILEM, The International Union of Testing and Research Laboratories for Materials and Structures, is an international, non-governmental technical association whose vocation is to contribute to progress in the construction sciences, techniques and industries, essentially by means of the communication it fosters between research and practice. RILEM activity therefore aims at developing the knowledge of properties of materials and performance of structures, at defining the means for their assessment in laboratory and service conditions, and at unifying measurement and testing methods used with this objective.

RILEM was founded in 1947, and has a membership of over 900 in some 80 countries. It forms an institutional framework for cooperation by experts to:

- optimize and harmonize test methods for measuring properties and performance of building and civil engineering materials and structures under laboratory and service environments;
- prepare technical recommendations for testing methods;
- prepare state-of-the-art reports to identify further research needs.

RILEM members include the leading building research and testing laboratories around the world, industrial research, manufacturing and contracting interests as well as a significant number of individual members, from industry and universities. RILEM's focus is on construction materials and their use in buildings and civil engineering structures, covering all phases of the building process from manufacture to use and recycling of materials.

RILEM meets these objectives though the work of its technical committees. Symposia, workshops and seminars are organized to facilitate the exchange of information and dissemination of knowledge. RILEM's primary output are technical recommendations. RILEM also publishes the journal *Materials and Structures* which provides a further avenue for reporting the work of its committees. Details are given below. Many other publications, in the form of reports, monographs, symposia and workshop proceedings, are produced.

Details of RILEM membership may be obtained from RILEM, École Normale Supérieure, Pavillon du Crous, 61, avenue du Pdt Wilson, 94235 Cachan Cedex, France.

RILEM Reports, Proceedings and other publications are listed below. Full details may be obtained from E & F N Spon, 2–6 Boundary Row, London SE1 8HN, Tel: (0)171-865 0066, Fax: (0)171-522 9623.

Materials and Structures

RILEM's journal, *Materials and Structures*, is published by E & F N Spon on behalf of RILEM. The journal was founded in 1968, and is a leading journal of record for current research,in the properties and performance of building materials and structures, standardization of test methods, and the application of research results to the structural use of materials in building and civil engineering applications.

The papers are selected by an international Editorial Committee to conform with the highest research standards. As well as submitted papers from research and industry, the Journal publishes Reports and Recommendations prepared buy RILEM Technical Committees, together with news of other RILEM activities.

Materials and Structures is published ten times a year (ISSN 0025-5432) and sample copy requests and subscription enquiries should be sent to: E & F N Spon, 2–6 Boundary Row, London SE1 8HN, Tel: (0)171-865 0066, Fax: (0)171-522 9623; or Journals Promotion Department, Chapman & Hall Inc, One Penn Plaza, 41st Floor, New York, NY 10119, USA, Tel: (212) 564 1060, Fax: (212) 564 1505.

RILEM reports

1 Soiling and Cleaning of Building Facades
2 Corrosion of Steel in Concrete
3 Fracture Mechanics of Concrete Structures – From Theory to Applications
4 Geomembranes – Identification and Performance Testing
5 Fracture Mechanics Test Methods for Concrete
6 Recycling of Demolished Concrete and Masonry
7 Fly Ash in Concrete – Properties and Performance
8 Creep in Timber Structures
9 Disaster Planning, Structural Assessment, Demolition and Recycling
10 Application of Admixtures in Concrete
11 Interfacial Transition Zone in Concrete
12 Performance Criteria for Concrete Durability
13 Ice and Construction

RILEM proceedings

RILEM recommendations and recommended practice